Jerzy Kowalewski

# Capacity Enhancement by Pattern-Reconfigurable Multiple Antenna Systems in Vehicular Applications

Karlsruher Forschungsberichte
aus dem Institut für Hochfrequenztechnik und Elektronik

*Herausgeber: Prof. Dr.-Ing. Thomas Zwick*

**Band 95**

Eine Übersicht aller bisher in dieser Schriftenreihe erschienenen Bände finden Sie am Ende des Buches.

# Capacity Enhancement by Pattern-Reconfigurable Multiple Antenna Systems in Vehicular Applications

by
Jerzy Kowalewski

Scientific
Publishing

Karlsruher Institut für Technologie
Institut für Hochfrequenztechnik und Elektronik

Capacity Enhancement by Pattern-Reconfigurable Multiple Antenna Systems in Vehicular Applications

Zur Erlangung des akademischen Grades eines Doktor-Ingenieurs von der KIT-Fakultät für Elektrotechnik und Informationstechnik des Karlsruher Instituts für Technologie (KIT) genehmigte Dissertation

von M. Sc. Jerzy Kowalewski

Tag der mündlichen Prüfung: 26. September 2019
Referent: Prof. Dr.-Ing. Thomas Zwick
Korreferent: Prof. Dr.-Ing. Leonardo Lizzi

**Impressum**

KIT Scientific Publishing

Karlsruher Institut für Technologie (KIT)
KIT Scientific Publishing
Straße am Forum 2
D-76131 Karlsruhe

KIT Scientific Publishing is a registered trademark of Karlsruhe Institute of Technology.
Reprint using the book cover is not allowed.

www.ksp.kit.edu



Print on Demand 2020 – Gedruckt auf FSC-zertifiziertem Papier

ISSN 1868-4696
ISBN 978-3-7315-0997-4
DOI 10.5445/KSP/1000099791

# Editor's Foreword

Mobile communication has become an important and for many of us an indispensable part of our society. The pure voice transmission is now accompanied by the data transmission, which enables new application possibilities for mobile equipment. The most important technology drivers in mobile communication include an enhancement of the available data rate or the channel capacity of the mobile communication system as well as extensive network coverage. A recent application of mobile communication is the vision of autonomous driving, which inevitably requires a stable and secure vehicular network. Since the bandwidth available for this purpose is severely limited, the most important option left for increasing the data rate (or the channel capacity) as well as the network coverage, that has recently become the focus of research and industry, is the concept of multiple-antenna systems. The classical diversity-systems will soon be replaced by MIMO (Multiple-Input-Multiple-Output) systems, which specifically employ the multipath characteristics of the radio channel for increasing the channel capacity. In a MIMO system, multiple bitstreams are spatially separated and independently transmitted by means of sub-channels, which enhances the overall channel capacity. The maximum channel capacity of a MIMO system is crucially dependent on an optimum antenna configuration at the transmitter and receiver end. As MIMO uses the multipath characteristics of the radio channel, an optimum-capacity antenna configuration should be adapted to the multipath radio channel. In contrast with classical systems, in which an antenna is developed on the basis of a given radiation pattern, a method is required, in which the antenna systems for the specified radio channels can be optimized in terms of channel capacity. That is where the work of Dr.-Ing. Jerzy Kowalewski starts.

In his dissertation, Jerzy Kowalewski has developed important scientific principles for capacity enhancement by pattern-reconfigurable multiple antenna systems in vehicular applications. He developed a novel method to find capacity maximizing radiation patterns for pattern-reconfigurable multiple antenna systems based on extensive channel simulations. He successfully verified his method by measurements in an urban environment. Dr. Kowalewski designed and realized several pattern-reconfigurable multiple antenna systems based on the design goals, that were he determined from his new method. Based on these antenna systems, he was able to demonstrate in his measurement campaigns a substantial capacity gain at a lower system complexity compared to the state-of-the-art antenna systems. His new antenna systems show a very good performance and provide a better form factor than the state-of-the-art antennas, which is very important in vehicular applications. Additionally, in cooperation with the University of Vienna, he was able to show that his antennas can be integrated into a cavity in carbon fiber, thus being completely hidden inside the roof.

Dr. Jerzy Kowalewski has demonstrated how to design a reconfigurable antenna for maximum capacity in a given radio channel. I am sure that his work represents an important innovation to the state-of-the-art. I am sure that his novel measurement methods will draw much attention and find many users worldwide. For Dr. Kowalewski, with his creativity and great knowledge of antennas, I wish him further much success in his scientific engineering career and economic endeavors.

Prof. Dr.-Ing. Thomas Zwick

- Institute Director -

# Abstract

The modern trends like internet of things (IoT) and automation penetrate different areas of our life. One of them is transportation and so the vehicles are equipped with numerous sensors, such as radio detection and ranging (radar), light detection and ranging (lidar), cameras etc. The data collected by these sensors is essential for autonomous driving. Yet, this data can be exchanged between the road users to enable e.g. safe maneuvering or saved for further analysis. This exchange is however not possible without reliable wireless links. For this purpose either dedicated systems like vehicle-to-vehicle (V2V) and vehicle-to-everything (V2X) or communication standards like 4G and 5G can be used. Antennas are inevitable to establish these wireless links. To fulfill the strict requirements on bandwidth and flexibility, and increase systems' performance, optimized antennas are required.

Considering the automotive case, the antennas change their orientation mostly in the azimuth plane. Therefore, unlike the isotropic channel in case of mobile phones, the automotive channel shows strong direction selectivity. It means that the channel parameters should be considered during both the design and the evaluation process. This work presents an antenna design methodology which includes the channel knowledge. The required data is acquired by means of ray tracing (RT) simulation and proved by a test drive in an urban environment. The proposed procedure can successfully be applied for arbitrary antenna systems on mobile platforms. Following the proposed design steps, relevant beam directions are identified prior to the design of the antenna structure. Different pattern reconfiguration techniques, to realize the antennas covering these directions, are studied and discussed in detail. All of them are tested by measurements of fabricated prototypes. Furthermore, two types of electronic switches, required to electronically change between the reconfigurable states, are compared. Based on this knowledge several reconfigurable multiple antenna systems are designed, fabricated and tested.

The performance of the proposed multiple antenna systems is first checked and compared to a conventional system, using omnidirectional antennas, by means of a channel based envelope correlation coefficient (ECC) calculation. Already this results show the superiority of the proposed antenna systems against conventional ones. The final proof is given by the channel capacity calculation from both virtual and real world test drives. The results confirm that applying the proposed methodology, during the design of pattern reconfigurable antenna system, leads to capacity enhancement of multiple antenna systems. The proposed antenna system improves the channel capacity by more than a factor of 2 compared to a conventional system without increasing the system's volume and number of front ends.

# Zusammenfassung

Moderne Trends wie Internet der Dinge (IoT) und Automatisierung durchdringen verschiedene Bereiche unseres Lebens. Einer von diesen Bereichen ist die Mobilität. Daher werden Fahrzeuge mit zahlreichen Sensoren ausgestattet, zum Beispiel für funkgestützte Ortung und Abstandsmessung (Radar), lichtgestützte Ortung und Abstandsmessung (Lidar), Kameras usw. Die von diesen Sensoren erfassten Daten sind für das autonome Fahren von wesentlicher Bedeutung. Diese Informationen können zwischen den Verkehrsteilnehmern ausgetauscht werden um z.B. sicheres Manövrieren zu ermöglichen. Dieser Austausch ist jedoch ohne zuverlässige drahtlose Verbindungen nicht möglich. Für diesen Zweck können entweder dedizierte Systeme wie Fahrzeug-zu-Fahrzeug (V2V) und Fahrzeug-zu-X (V2X) oder Kommunikationsstandards wie 4G und 5G vorgesehen werden. Antennen sind unvermeidlich um diese drahtlosen Verbindungen herzustellen. Um die strengen Anforderungen an Bandbreite und Flexibilität zu erfüllen und die Systemperformanz zu steigern sind optimierte Antennen erforderlich.

Im Fall der Automobilantennen ändern die Antennen ihre Ausrichtung meist nur in der Azimut Ebene. Im Gegensatz zum isotropen Kanal bei Mobiltelefonen, weist der Automobilkanal eine starke Richtungsselektivität auf. Dies bedeutet, dass die Kanalparameter sowohl während des Entwurfs als auch der Evaluierung berücksichtigt werden müssen. Diese Arbeit präsentiert eine Antennenentwurfsmethodik, die das Kanalwissen miteinschließt. Die erforderlichen Daten werden mittels Ray Tracing (RT) Simulationen erfasst und durch eine Probefahrt in einer städtischen Umgebung verifiziert. Das vorgeschlagene Verfahren kann für beliebige Antennensysteme auf mobile Plattformen angewendet werden. Entsprechend den vorgeschlagenen Entwurfsschritten, werden vor dem Entwurf der Antennenstruktur die relevanten Strahlrichtungen identifiziert. Anschließend, werden verschiedene Patternrekonfigurationstechniken eingehend studiert und diskutiert. Diese Techniken ermöglichen die Fertigung von Antennen, die die gewünschten Richtungen abdecken. Diese Antennen werden durch

Aufbau und Messungen von Prototypen getestet. Außerdem werden zwei Arten von elektronischen Schaltern verglichen, die für die elektronische Umschaltung zwischen rekonfigurierbaren Zuständen erforderlich sind. Basierend auf diesem Wissen werden mehrere rekonfigurierbare Mehrantennensysteme konstruiert, gefertigt und verifiziert.

Die Leistungsfähigkeit der vorgeschlagenen Mehrantennensysteme wird zuerst mittels eines kanalbasierten Hüllkurvenkorrelationskoeffizienten (ECC) Berechnung evaluiert und mit einem herkömmlichen System aus Rundstrahlantennen verglichen. Bereits diese Ergebnisse zeigen die Überlegenheit der vorgeschlagenen Antennensysteme gegenüber den konventionellen Systemen. Schließlich wird die Kanalkapazität basierend auf Ergebnissen von virtueller Probefahrten und einer Messkampagne in städtischer Umgebung berechnet. Die Ergebnisse bestätigen, dass die Anwendung der vorgeschlagenen Methodik zur Kapazitätssteigerung von Mehrantennensysteme führt. Das vorgeschlagene Mehrantennensystem steigert die Kanalkapazität um mehr als Faktor zwei ohne den Hardwareaufwand wie auch Systemabmessung zu erhöhen.

# Acknowledgment

I would like to thank Prof. Dr.-Ing. Thomas Zwick for giving me the opportunity to work on this topic and supervising this work. I am personally indebted to Ronald Vester for etching of the antenna structures. My acknowledgement goes to Dr.-Ing. Daniel Müller and Christian Bohn for proofreading. I would like especially thank Jonathan Mayer for refreshing and inspiring discussions and proofreading. For support in the channel measurement campaign, I would like to thank my colleague Jörg Eisenbeis as well as Dr. Zsolt Kollar and Magnus Tingulstad. A special thanks go to Tobias Mahler with whom I have shared office and discussed the technical issues for long years. I am grateful to Lukasz Zwirello who have always supported my with valuable hints concerning Ph. D. thesis as well as life in Karlsruhe. Furthermore, I would like to thank all colleagues at IHE for their friendly and supportive attitude, which made my time at IHE an important experience in my life. Further thanks go to Andreas Gallego, Mirko Nonnenmacher and Andreas Lipp, technical staff of the IHE, for their support in realization of the projects. Moreover, I would like to thank Marion Jentzsch, Simone Gorre und Angela Ziemba for their support with bureaucratic problems and business trips organization. I am very grateful to my parents who have always encouraged me in my decisions and supported me in coming to Karlsruhe for my master studies. Finally, my greatest thanks goes to my wife Fanny for her support and forbearance for my work.

# Contents

# Abbreviations and Symbols

## Abbreviations

| | |
|---|---|
| **ACF** | Autocorrelation Function |
| **AM** | Amplitude Modulation |
| **AoA** | Angle of Arrival |
| **AoD** | Angle of Departure |
| **APS** | Angular Power Spectrum |
| **BS** | Base Station |
| **BW** | Band Width |
| **CAD** | Computer Aided Design |
| **CDF** | Cumulative Distribution Function |
| **CFRP** | Carbon Fiber Reinforced Polymer |
| **CMOS** | Complementary Metal-Oxide-Semiconductor |
| **CPW** | Coplanar Waveguide |
| **DAB** | Digital Audio Broadcasting |
| **DC** | Direct Current |
| **DVB-T** | Digital Video Broadcasting–Terrestrial |
| **ECC** | Envelope Correlation Coefficient |
| **ETC** | Electronic Toll Collection |

| | |
|---|---|
| **ESPAR** | Electronically Steerable Passive Array Radiator |
| **FBR** | Front-to-Back Ratio |
| **FDTD** | Finite Difference Time Domain |
| **FEM** | Finite Element Method |
| **FFT** | Fast Fourier Transform |
| **FM** | Frequency Modulation |
| **FOM** | Figure of Merit |
| **FSPL** | Free Space Path Loss |
| **GO** | Geometrical Optics |
| **GPS** | Global Positioning System |
| **GSM** | Global System for Mobile Communications |
| **HPBW** | Half Power Beamwidth |
| **IFA** | Inverted F Antenna |
| **IHE** | Institute of Radio Frequency Engineering and Electronics |
| **IoT** | Internet of Things |
| **KIT** | Karlsruhe Institute of Technology |
| **LNA** | Low Noise Amplifier |
| **LOS** | Line of Sight |
| **LTE** | Long Term Evolution |
| **MEMS** | Micro-Electro-Mechanical System |
| **MIMO** | Multiple-Input Multiple-Output |
| **MPC** | Multipath Component |
| **OFDM** | Orthogonal Frequency Division Multiplexing |

| | |
|---|---|
| **OSM** | Open Street Map |
| **PCB** | Printed Circuit Board |
| **PDP** | Power Delay Profile |
| **RF** | Radio Frequency |
| **RFID** | Radio Frequency Identification |
| **RT** | Ray Tracer |
| **Rx** | Receiver |
| **QFN** | Quad Flat No Leads |
| **SDARS** | Satellite Digital Audio Radio Services |
| **SDR** | Software Defined Radio |
| **SGP** | Slotted Ground Plane |
| **SMA** | SubMiniature version A |
| **SNR** | Signal-to-Noise Ratio |
| **TPMS** | Tyre Pressure Monitoring System |
| **Tx** | Transmitter |
| **UE** | User Equipment |
| **UMTS** | Universal Mobile Telecommunications Service |
| **USRP** | Universal Software Radio Peripheral |
| **V2V** | Vehicle-to-Vehicle |
| **V2X** | Vehicle-to-Everything |
| **WiFi** | Wireless Internet |

# Constants

| | |
|---|---|
| $\pi$ | Pi: 3.14159 . . . |
| $c$ | speed of light: 299 792 458 m/s |
| $k$ | Boltzmann constant: $1.380\,649 \times 10^{-23}$ J/K |

# Latin Symbols and Variables

## Small letters

| | |
|---|---|
| $d$ | distance |
| $d_{Rx}$ | Rx antenna spacing |
| $d_{Tx}$ | Tx antenna spacing |
| $f$ | frequency |
| $h$ | impulse response |
| $n$ | number of response |
| $t$ | time |
| $x(t)$ | input signal |
| $y(t)$ | output signal |

## Capital letters

| | |
|---|---|
| $B$ | bandwidth |
| $B_c$ | coherence bandwidth |
| $C$ | channel capacity |
| $C(\theta, \psi)$ | radiation pattern |
| $D$ | antenna directivity |
| $E$ | electric field strength |
| $H$ | channel matrix |
| $I$ | identity matrix |
| $M$ | number of transmit antennas |
| $N$ | number of receive antennas |
| $N_{FFT}$ | FFT length |
| $P(\theta, \psi)$ | complex angular power spectrum |
| $P_{Tx}$ | power at the transmitter |
| $S$ | transmitted signal power |

| | |
|---|---|
| $T$ | temperature |
| $T_c$ | coherence time |
| $X$ | input signal matrix |
| $Y$ | output signal matrix |

## Greek Symbols and Variables

| | |
|---|---|
| $\varepsilon$ | electric permittivity |
| $\varepsilon_0$ | free space permittivity |
| $\varepsilon_r$ | relative permittivity |
| $\mu$ | magnetic permeability |
| $\mu_0$ | free space permeability |
| $\mu_r$ | relative permeability |
| $\eta$ | efficiency |
| $\theta$ | elevation angle |
| $\psi$ | azimuth angle |
| $\lambda$ | wavelength |
| $\lambda_i$ | i eigenvalue |
| $\sigma$ | standard deviation |
| $\varphi$ | phase shift |

## Operators and mathematical Symbols

| | |
|---|---|
| $\underline{a}$ | complex number |
| $\vec{\underline{a}}$ | complex vector |
| $a^*$ | complex conjugate |
| $a^\dagger$ | complex conjugate transpose |
| $\overline{a}$ | mean value |
| $\boldsymbol{I}_N$ | identity matrix of dimension $N$ |
| $\circ\!\!-\!\!\bullet$ | Fourier transform |
| $\bullet\!\!-\!\!\circ$ | inverse Fourier transform |
| $\propto$ | proportional |

# Common Subscripts

| | |
|---|---|
| $c$ | center |
| $max$ | maximal |
| $r$ | relative |
| $Rx$ | receiver |
| $Tx$ | transmitter |

# 1 Introduction

Within the last decade the microwave technology experienced rapid development and has spread into various domains of our life. Radar and localization systems support the users in their everyday tasks, while communication systems, such as global system for mobile communication (GSM), universal mobile telecommunications service (UMTS) and long term evelution (LTE) have sustainably changed the world. When radar systems are present in various domains, wireless communication have dominated nearly every domain of people's life. Recently the concept of internet of things (IoT), a global system which enables electronic devices, sensors and actuators to connect and exchange data, was developed. In order to support the furthe development of the IoT the wireless communication systems are essential. Already today many mobile devices like notebooks, tablets and smartphones take advantage of this technology to keep users connected to the global network. However, the concept of IoT does not end here, it expands to smart home and smart factory. Since the users demand to stay connected all the time also cars are now equipped with modern wireless communication systems, providing the users a seamless transition from the smartphone to the vehicle. At the moment they are used mainly to inform users about traffic situations, however this is the first important step towards vast data exchange and finally autonomous driving. The cars are thus no longer just mechanical vehicles bringing us from point A to point B, but they are dynamic part of the IoT network and have the ability to exchange the crucial data and use the roads more safely and efficiently [DBG+10].

## 1.1 Motivation

Recently the development towards ubiquitous use of multiple wireless services could be observed in the automobile industry. Modern cars are equipped with antennas for various wireless standards [PSR+11]:

- amplitude modulation / frequency modulation (AM/FM) radio
- digital audio broadcasting (DAB) and digital video broadcasting (DVB-T)
- global system for mobile communications (GSM)
- long term evolution (LTE)
- vehicle-to-vehicle (V2V) and vehicle-to-everything (V2X) communication
- wireless internet (Wi-Fi)
- satellite navigation systems (GPS)
- satellite digital radio (SDARS)
- remote keyless entry
- tyre pressure monitoring system (TPMS)
- electronic toll collection (ETC)

Additionally to these wireless services there are also different radar systems mounted on the car. All these systems use the low kHz range up to the millimeter wave frequency range.

Almost all of these systems need separate antennas, and thus the number of antennas mounted on modern cars is growing. At the same time the aesthetic and aerodynamics expectation is that they stay invisible. While antennas for radio and video broadcasting systems are no longer mounted on the car roof, but are nowadays integrated in the glass of the windows, the antennas for communication and navigation systems are still mounted on the roof inside a so-called shark-fin [PSR+11]. Despite the advantage of a wide view angle, this position possesses certain problems for antenna designers. The shark-fin housing offers only limited space and has to stay low profile, to not increase the aerodynamic drag. Alternatively the antennas can be mounted in the side mirror housing, however this location also offers only limited space and the view range is smaller than in the case of the car roof. Whichever antenna mounting position is selected, the antennas has to be as compact as possible.

The trend in the automotive industry is to equip vehicles with growing number of integrated sensors. Thus the cars become more aware of their environment. In order to use the collected data the interaction between vehicles, other users and the infrastructure is essential. For this interaction either dedicated systems like V2V or V2X or existing and future communication standards like 4G and 5G can be used [MMG17, FDAZ+18]. At the same time the 4G and 5G systems are used for infotainment systems, thus high data throughput, reduced end-to-end latency and improved reliability are needed. It is important to remember that the vehicles are constantly changing their driving direction and speed between 0 and 250 km/h. Also their environment, such as urban, rural, highway, undergoes rapid changes like e.g. number of surronding vehicles or vegetation density. These factors influence the mobile channel severely and can cause connection interruptions e.g. due to fading. In 5G it is forseen to use massive multiple-input multiple-output (MIMO) antenna systems [Lar17] at the base station (BS), thus having large number of antennas to overcome the problems of fading and support tens of users simultaneously. Nevertheless it is not possible to mount massive MIMO antennas in the shark-fin housing, since the space inside these modules is already limited. In this case pattern reconfigurable MIMO antennas are an attractive solution.

Yet, it is not enough to just design a MIMO antenna system for V2X communication. The performance of the antenna should be measured. Evaluating the performance of the system only in terms of matching, gain and radiation pattern is however not sufficient. The dynamic behavior and channel capacity has to be evaluated under real conditions by test drives in different environments such as city, countryside and highway as soon as the prototype is fabricated [ETKM13]. Since the antenna systems are developed for new car models not yet available on the market, secrecy is needed and the vehicles are covered with camouflage, making the preparation for the test drives time consuming. In addition, test drives are not reproducible and involve a great financial expenditure. Another measure used very often for MIMO antennas is envelope correlation coefficient (ECC). It is a very good figure of merit (FOM), nonetheless the channel is not taken into consideration, an thus the distribution of the field impinging at the antenna is ignored. Therefore it is important to find a method that gives a more accurate measure of an antenna system's performance.

## 1.2 State of the Art

The antenna is an interface between the channel and the transceiver and is thus essential for proper function of every wireless system. Apart from requirements on bandwith and matching, the important question in case of an automotive communication antenna is what should be its gain and radiation characteristic. Nowadays antennas with radiation patterns omnidirectional in azimuth are used [GL14, GL15, EPKM14]. Considering the fact that the car is moving and changing its driving direction while the antenna is placed on the car roof, omnidirectional pattern seems to be the obvious choice, since signals can impinge the car from all angular directions. This logic is true for line of sight (LOS) connection, however the situation changes for multipath rich urban environments (see Fig. 1.1), where LOS is very often interrupted due to shadowing. In this case the use of antennas with directive patterns leads to better results, as presented by Reichardt et al. [RMSZ13]. It is shown that due to existence of dominant angles of arrival (AoA) different radiation patterns are advantageous for urban and rural scenarios. In order to establish which radiation patterns are advantageous, an antenna synthesis procedure using ray-tracing simulations for multiple antenna systems as presented in [MRH+15] can be used. Resulting from this procedure are near-optimum vehicular MIMO radiation patterns. The patterns enabling the increased channel capacity should be directed towards front and rear of the vehicle and the other ones orthogonal to it (towards left and right side of the vehicle).

Using directive antennas covering the named angular directions will most likely improve the system performance in many situations, however this might also result in lower performance for certain other scenarios. Therefore it is advantageous to use a more flexible solution like pattern reconfigurable antennas. In this case the patterns can be switched dynamically while the car is moving and the best pattern can be chosen, resulting in an increased reliability.

Within recent years the topic of pattern reconfigurable antennas was thoroughly researched and some techniques to realize them were proposed. In [NJDP12] switching between elements pointing in different directions is used for pattern reconfiguration. Such solution offers good flexibility, however is not compact enough for automotive communication (footprint of this design is $0.54\lambda_0 x 0.54\lambda_0$). An interesting posibility are multimode antennas with pattern selection [KL13, SVM15, ZWZ+15]. However, while these designs are compact,

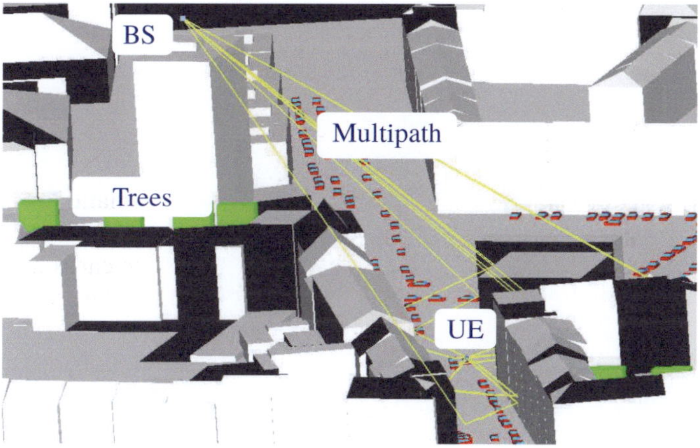


Figure 1.1: A picture from RT tool presenting multipath propagation in an urban scenario. BS: base station, UE: user equipment (vehicle).

their flexibility in terms of pattern direction is limited, since only beams resulting from antenna's characteristic modes are possible. An interesting approach are electronically phase-tunable array antennas [ME16, ZNTF18]. These antennas consist of multiple elements and the signal phase at the elements can be controlled with help of varactors. A very popular solution is the use of parasitic elements [GO00, ORPH+17], where the beam direction can be changed by activating parasitic elements in vicinity of the primary radiator. This approach offers a high flexibility (arbitrary number of reconfigurable patterns) and relatively compact construction at the same time. Another very interesting concept was introduced and is being further developed by the group of Prof. Fumeaux [NTKHF15, NTPHF17a, NTPHF17b]. They proposed reconfigurable substrate-integrated cavities. This solution is very low profile but has limited bandwidth (around 2 % fractional bandwidth) and low efficiency (simulated below 60 %).

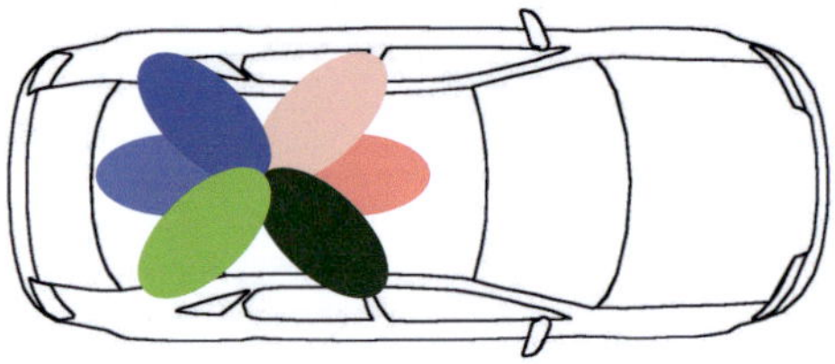

Figure 1.2: Directions that should be covered by automotive antenna based on results of an urban channel simulation.

## 1.3 Problem Description

If we consider that vehicles can change their direction in one plane only unlike e.g. mobile phones changing their orientation in all three dimensions, the automotive wireless channel gets directional, instead of isotropic. Based on this knowledge we can improve the design process of the antenna by adding radiation pattern requirements to the list of initial parameters. The information about the channel's directivity can be obtained from RT simulations in form of an angular power spectrum (APS).

This process and the obtained results are explained in detail in section 2.8. Based on these results, the beams should cover front, front-left and front-right direction, as well as rear, rear-left and rear-right direction with regard to the vehicle (see Fig. 1.2). These patterns are defined for a car-roof antenna, since this mounting position offers the best field of view and an elevated antenna placement. The graphical representation of channel simulation in figure 1.3 shows that the waves propagating from the base station diffract on roof edges of the buildings or get reflected on the obstacles and get trapped in the street canyon. Walls of the buildings limiting the street establish a waveguide for the propagating waves. As a result the rays undergo multiple bounces before they reach the antennas on vehicle's roof. This phenomena is called canyoning effect and explains the mentioned beam directions.

If we consider the multipath rich urban environment and the direction selective nature of the channel we can assume that applying a MIMO system and beamforming will result in a channel capacity improvement. Knowing this we can

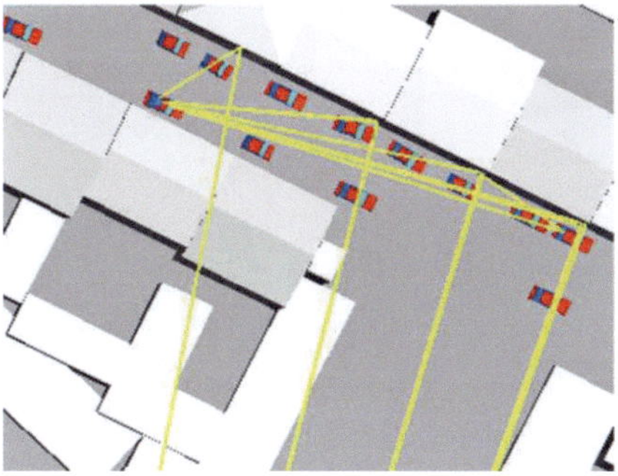

Figure 1.3: Picture from the RT tool presenting canyoning effect in an urban environment. Yellow lines represent signal paths.

assume that replacing conventional omnidirectional antennas with directional ones pointing in proper directions will result in a signal-to-noise-ratio (SNR) improvement. Furthermore, if some of the beamforming effort is moved to the antenna by utilizing pattern reconfigurable antennas, the capacity improvement resulting from spatial multiplexing can be even higher. Another aspect is the volume available for antennas within the shark-fin module. Reducing the number of antennas and antenna size is crucial for vehicular applications, since only limited space is available. Thus, a multiple antenna system utilizing pattern reconfigurable antennas can contribute to channel capacity enhancement of the overall system without substantial increase in cost and volume of the antenna.

## 1.4 Design Methodology

In order to design a multiple antenna system for capacity enhancement in vehicular applications a proper design and test procedure is needed. A flow chart diagram of the design methodology proposed in this work is presented in Fig. 1.4. A key element in this procedure is the channel simulation. As already mentioned in the previous section, the information about the channel's direction selectivity can be obtained from the simulation results. Thus, the directions covered by radiation patterns can be defined as requirements for antenna systems prior to the design. The next step after defining the requirements is the choice of an

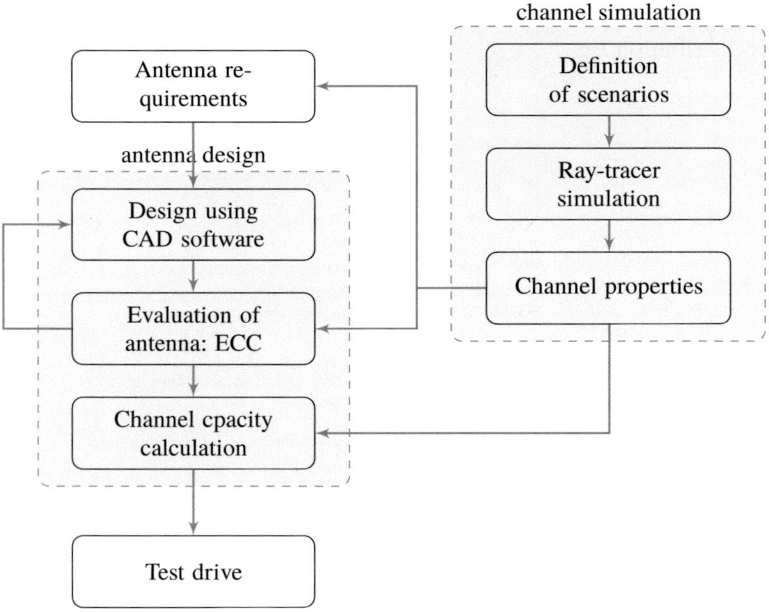


Figure 1.4: Flow chart presenting the design methodology proposed in this work.

appropriate antenna concept and design supported by the use of computer-aided design (CAD) software. The optimization using CAD software allows to improve the parameters like antenna matching, port isolation, gain, pattern direction and front-to-back ratio (FBR). However, these measures of the antenna's performance do not give us the ultimate answer if the antenna system is going to perform well in the given application. Therefore, a more systematic measure including knowledge about the channel is needed.

Channel capacity is a very good performance measure of an antenna system and it is sufficient enough to reliably compare two different designs. However, using the channel capacity to improve one design is cumbersome as a direct connection to radiation patterns is missing. Therefore, it is of advantage to first evaluate the design using ECC including information about the channel. The influence of the channel is given by the incident field at the antenna position, this information is obtained from RT simulation. The incident field is described by an angular power spectrum $P(\theta, \psi)$. In a standard ECC calculation a uniform power distribution $P_{\text{Uniform}}(\theta, \psi) \propto 1$ is assumed.

In order to evaluate antenna's performance averaged over given propagation scenarios, the ECC given by [Kil15]:

$$\rho_e = \frac{|\iint\limits_{4\pi} \overrightarrow{C_1}(\theta, \psi) \cdot \overrightarrow{C_2}^*(\theta, \psi) d\Omega|^2}{\iint\limits_{4\pi} |\overrightarrow{C_1}(\theta, \psi)|^2 d\Omega \iint\limits_{4\pi} |\overrightarrow{C_2}(\theta, \psi)|^2 d\Omega} \tag{1.1}$$

is adopted and completed by including the APS in the equation.

$$\rho_e = \frac{|\iint\limits_{4\pi} \overrightarrow{C_1}(\theta, \psi) P(\theta, \psi) \cdot \overrightarrow{C_2}^*(\theta, \psi) P(\theta, \psi) d\Omega|^2}{\iint\limits_{4\pi} |\overrightarrow{C_1}(\theta, \psi) P(\theta, \psi)|^2 d\Omega \iint\limits_{4\pi} |\overrightarrow{C_2}(\theta, \psi) P(\theta, \psi)|^2 d\Omega} \tag{1.2}$$

$\overrightarrow{C_n}(\theta, \psi)$ denotes the complex radiation pattern of the antenna $n$ and $d\Omega$ denotes the differential solid angle.

Based on the obtained results a redesign and optimization of the design can be conducted and the overall antenna system can once again be evaluated.

As soon as an optimized design is available, the channel capacity of the system can be calculated based on the simulation. Even though a simulation cannot possibly capture all the effects influencing the performance in a fully functional device, its value is indisputable in evaluating the design in the early phases of the process. Furthermore, the simulation is cheaper and less time consuming than a test drive [ETKM13]. Since we can control the environment in the simulation, it is more reliable to compare the simulated performance of two different designs than to do that based on measurement campaign in dynamically changing environment. Nevertheless, the final prototype should be implemented in the vehicle and as a last step test drives can be performed.

## 1.5 Goals and Organization of the Work

The goal of this work is to define a design and test procedure for capacity enhancement in automotive antenna systems. Pattern reconfigurable antenna systems are proposed as a solution and investigated in this work. Therefore, different possible approaches for the design of pattern reconfigurable antennas realizing identified radiation patterns are examined.

Requirements defined for antennas developed within the scope of this work is a bandwidth of at least 200 MHz, which is sufficient enough for mobile communication. The center frequency should placed between 2.45 GHz and 2.6 GHz. Therefore, outdoor measurements in unlicensed band are possible or the antennas can be directly applied in the LTE bands 40 and 41 if 2.6 GHz is chosen.

Chapter 2 gives us an insight into wireless channel parameters. Furthermore, the simulation procedure using the ray tracer tool is explained in detail. Finally, the patterns optimized for automotive communication applications are determined using the RT tool.

Chapter 3 presents the design of pattern reconfigurable antennas. Different techniques are shown and discussed. Among them element switching, utilizing parasitic elements and phase switching seem to be most promising, and are tested by design and fabrication of antenna models. Additionally, two different switching devices: p-i-n diodes and MEMS (micro-electro-mechanical system) are evaluated.

In chapter 4 the findings of the previous chapters are combined in the design of pattern reconfigurable multiple antenna systems. Four different designs meeting the defined requirements are realized and tested.

Chapter 5 presents the evaluation methods proposed for testing of pattern reconfigurable multiple antenna systems. The channel based envelope correlation coefficient (ECC) is introduced as a tool for verification during optimization process. Channel capacity calculation from simulated and measured values collected during virtual and real world test drives are then used as a final measure of antenna systems' performance.

# 2 Wireless Channel

In case of wireless communication, the transmitting and receiving antenna and the wireless channel replace the transmission cable. The transmitting antenna converts a guided electromagnetic wave into electromagnetic waves capable of propagating in the free-space, whereas the receiving antenna converts the electric (or magnetic) field captured from the environment into alternating current [Bal05]. A theoretical isotropic radiator produces a wave propagating in all directions. This wave is attenuated depending on the distance from the source. This attenuation is known as free-space path loss (FSPL) and is given by formula [VBA03]:

$$\text{FSPL} = \left(\frac{4\pi d}{\lambda}\right)^2 \tag{2.1}$$

where $d$ is the distance from the source and $\lambda$ is the wavelength.

If the path of the wave between transmitter and receiver is undisturbed, a so-called line of sight connection is established. On the other hand if the wave encounters an obstacle, depending on its form and size, the following propagation mechanisms can be observed [VBA03]:

- **reflection** is the abrupt change in the direction of a wave impinging at a boundary between two different media. The size of the surface should be multiples of a wavelength.
- **diffraction** is the spreading of waves around the edges of an obstacle. It can be well explained by Huygens' principle which states that each point on a wavefront acts as a fresh wave source.
- **scattering** is the interaction of a wavefront with a rough surface or an object with a size smaller than the wavelength. In case of scattering the wave reflected from the object is spread over a solid angle range rather than one direction [GW98].

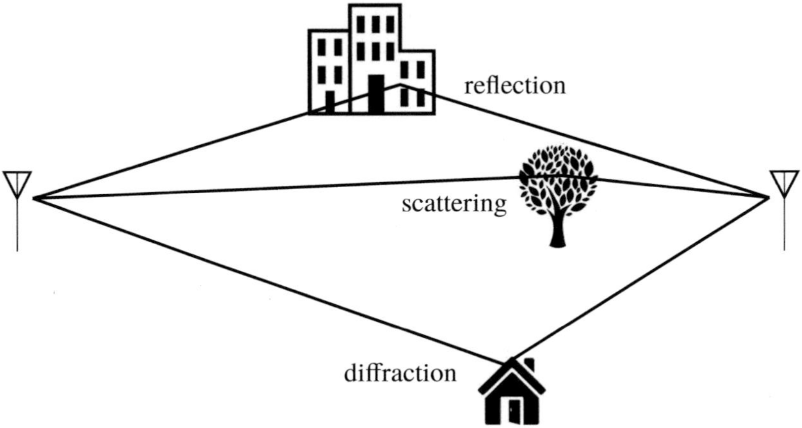


Figure 2.1: Schematic drawing of a wireless channel with different propagation phenomena.

Thus, a multipath propagation appears (see Fig. 2.1).

The multipath components superimpose at the receiver, and the amplitude and phase of the individual components are crucial in this process. This process is known as multipath interference and can either have positive or negative influence on the transmission, depending on the propagation environment. In some cases, under unfavorable conditions, destructive interference can be observed and thus the signal-to-noise ratio (SNR) at the receiver decreases. This effect is known as fading.

The antennas themselves influence the transmission through their directivity patterns, position and polarization. Therefore diversity or MIMO antenna systems can be used to overcome problems with fading, by exploiting multiple observations of the channel to improve the SNR [Bre59]. Furthermore, MIMO systems use the multipath propagation to their advantage by establishing multiple channels. In the following chapter a more detailed description of the wireless channel and MIMO systems is given. Moreover the important antenna parameters are discussed. Following, the utilized RT tool is introduced. Finally, the patterns required for automotive antenna systems are determined using this RT tool.

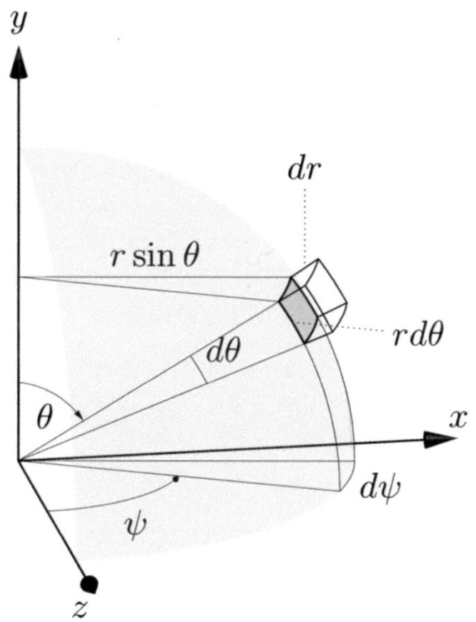


Figure 2.2: The spherical coordinate system used in this work. $r$ represents the radius, $\psi$ and $\theta$ represent azimuth and elevation angles respectively.

## 2.1 Antenna fundamentals

As already discussed at the beginning of this chapter, the antenna transforms a guided electromagnetic wave into a wave propagating in free-space, or vice verse. It means, it is an interface between the channel and the RF front-end. Therefore, it is crucial to understand the parameters characterizing the antenna before realizing an optimized system.

### 2.1.1 Key properties of antennas

In order to be able to characterize, evaluate and compare antennas, a set of parameters is used by antenna engineers. One of the most commonly used parameters is the bandwidth (BW). It refers to the frequency range within which an antenna characteristic fulfills certain requirements. The most often used is the impedance bandwidth, and it refers to the frequency range over which the return loss is lower than some given value. In case of wireless communication 6 dB return loss is considered sufficient (around a quarter of the power is reflected), however in case of radar systems 20 dB might be required. The gain bandwidth is also commonly used and it refers to a frequency range within which the gain exceeds some given value (e.g. half the value of maximum gain $BW_{3dB}$). Another important characteristics is the antenna's radiation pattern. This measure represents the amount of energy radiated by the antenna in each spatial direction. Most often it is presented graphically using a spherical coordinate system (see Fig. 2.2). It is defined as:

$$C(\theta, \psi) = \left. \frac{|\underline{\vec{E}}(\theta, \psi)|}{|\underline{\vec{E}}(\theta, \psi)|_{\max}} \right|_{r \to \infty} \tag{2.2}$$

where $C(\theta, \psi)$ represents the radiation pattern and $\underline{\vec{E}}(\theta, \psi)$ is the radiated E-field.

A further parameter used to characterize antennas is gain. It is the ratio of the surface power density in a given direction for a given antenna, compared to the surface power density resulting from isotropic radiator. Thus, gain is given in unit dBi. The surface power density of the isotropic radiator is given by [Bal05]:

$$|\vec{S}| = S_i = \frac{P_{\mathrm{Tx}}}{4\pi r^2} \tag{2.3}$$

Where $P_{\mathrm{Tx}}$ is the power fed to the lossless isotropic source. Thus, the gain can be defined as:

$$G = \eta D = \eta \frac{S_{r\,\max}}{S_i} = \eta 4\pi r^2 \frac{S_{r\,\max}}{P_{Tx}} \tag{2.4}$$

$D$ defines the antenna's directivity, while $\eta$ defines the antenna's efficiency, and its value is between 0 and 1. The antenna efficiency contains all losses of the antenna.

The antenna polarization defines the polarization of the wave radiated by the antenna. Normally it refers to the electric field vector. One can differentiate between linear and circular polarization. Most terrestrial systems use linearly polarized waves, while satellite tend to use circular polarization which is more robust to changes in the polarization happening in the ionosphere (e.g. GPS antennas).

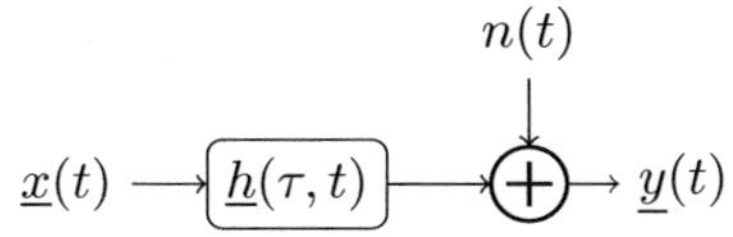


Figure 2.3: Mathematical representation of the interaction of the channel with the signal.

## 2.2 Channel description

A simplified model of the mathematical channel representation is shown in figure 2.3. The output signal $\underline{y}(t)$, which we get at the receiving antenna, depends on the transmission signal $\underline{x}(t)$ and the channel $\underline{h}(\tau, t)$ characterized by the impulse response (in case of a linear channel). It can be represented as a convolution of both terms in time domain. The convolution in time domain corresponds with a multiplication in the frequency domain.

$$\underline{y}(t) = \underline{h}(\tau, t) * \underline{x}(t) = \int_{-\infty}^{\infty} \underline{h}(\tau, t)\underline{x}(t-\tau)d\tau \tag{2.5}$$

$$\circ\!\!-\!\!\bullet$$

$$\underline{Y}(f) = \underline{H}(f) \cdot \underline{X}(f), \tag{2.6}$$

where $\underline{Y}$ is the output signal matrix, $\underline{X}$ is the input signal matrix and $\underline{H}$ is the channel matrix.

In order to account for random processes that occur in the channel, additive white Gaussian noise (AWGN) is added to the signal at the receiver (see Fig 2.3). It can be noted in frequency domain as:

$$\underline{Y}(f) = \underline{H}(f) \cdot \underline{X}(f) + \underline{N}(f) \tag{2.7}$$

The multiplicative noise is already accounted for in the channel impulse response [Sau99].

## 2.3 Time variance

All of the automotive communication scenarios considered in this work are dynamic and the cars within them travel with constant speed. Such moving objects lead to time variance in the channel i.e. the path lengths, propagation time and phase change over time. Furthermore, some paths might disappear and be replaced by some new paths. Thus, the result of superposition of multipath components at the receiver varies over time. Since the receiver changes the position in the channel, constructive and destructive interference are observed within the short distance of $\frac{\lambda}{2}$. This effect is called fast fading [Sau99].

In contrast to fast fading, slow fading originates from slow changes in the environment e.g. changing weather conditions. Additionally, while the transmitter or receiver move in their environment, they can be shadowed by a big obstacle such as a mountain or a block of buildings. In this case, unlike for fast fading where path length or phase of multipath components changes, the number and amplitude of the multipath components changes [GW98].

Since the radio channel considered in this work undergoes dynamic changes, one also has to consider the Doppler effect. It is a change of the apparent frequency of the impinging wave, which results from the object's movement [Sau99]. The change in frequency is proportional to the component of the object's speed in the direction of the wave and is given by the formula:

$$f_d = \frac{v}{\lambda} cos\alpha = f_c \frac{v}{c} cos\alpha = f_m cos\alpha, \tag{2.8}$$

where $f_d$ is the Doppler frequency, $f_c$ denotes the center frequency, $v$ is the object's speed and $\alpha$ is the angle between the speed vector and the wavefront.

If the multipath channel is considered, each individual wave arriving at the receiver from an arbitrary direction has its associated Doppler shift. As a result, the bandwidth of the received signal is spread compared to the transmitted bandwidth, thus the effect is known as Doppler spread. This effect can by analyzed in time domain by means of the coherence time. The coherence time $T_c$ describes the time over which the channel changes only slightly and the autocorrelation function is close to unity, thus the channel is time invariant.

The coherence time is inversely proportional to the Doppler spread of the channel [Sau99].

$$T_c \propto \frac{1}{f_m} \tag{2.9}$$

## 2.4 Frequency selectivity

In the dynamic channels considered in this work, the signal arriving at the receiver is composed of multipath components with noticeably different delays and different frequencies. The strength of the received signal at a given frequency $f_0$ is determined, among other things by the phase differences of the partial waves at $f_0$. Due to the different wavelengths, different signal paths and interaction points are observed for the individual frequencies. Thus, if the time of arrival for the received signals varies significantly, the channel varies with the frequency, which leads to a frequency-selective behavior of the channel.

The variation of the mean power with delay can be expressed graphically as different taps placed along the time axis and is know as power delay profile (PDP). It can be given with the formula:

$$P(t,\tau) = |\underline{h}(t,\tau)|^2. \tag{2.10}$$

The PDP can be characterized by:

- Mean delay $\mu_\tau(t)$

$$\mu_\tau(t) = \frac{\sum_{i=1}^{n} P_i \tau_i}{\sum_{i=1}^{n} P_i} \tag{2.11}$$

- Delay spread $\sigma_\tau(t)$

$$\sigma_\tau(t) = \sqrt{\frac{\sum_{i=1}^{n} P_i \tau_i^2}{\sum_{i=1}^{n} P_i} - \mu_\tau(t)^2}, \tag{2.12}$$

where $n$ is the number of samples.

The coherence bandwidth $B_c$ indicates the bandwidth over which the channel can be assumed to be almost constant, thus not frequency-selective. If the bandwidth $B$ of the communication system fulfills the requirement: $B \ll B_c$, the channel is frequency flat and no equalization is needed. The coherence bandwidth is inversely proportional to the delay spread:

$$B_c \propto \frac{1}{\sigma_\tau(t)}. \tag{2.13}$$

The coherence bandwidth can be determined via the frequency autocorrelation function (ACF) $\underline{r}^f_{HH}$ of the transfer function:

$$\underline{r}^f_{HH}(\Delta f, t) = \int_{-\infty}^{\infty} \left(\underline{H}^{LP}(f,t)\right)^* \underline{H}^{LP}(f+\Delta f, t)df, \tag{2.14}$$

where $\underline{H}^{LP}(f,t)$ is the equivalent low-pass transfer function. The spectrum at positive frequencies is shifted to the baseband ($f = 0$) and the amplitude is doubled. This equivalent transfer function simplifies the mathematical description [GW98].

In order to simplify the progression of the frequency ACF by describing it with a single parameter, the coherence bandwidth is defined as the frequency for which the magnitude of the correlation coefficient first drops below a predetermined value $e$ (i.e. $1/e$).

## 2.5 Direction selectivity

Another important issue of the wireless channel is the direction selectivity. The angle of departure (AoD) and the angle of arrival (AoA) of multipath components depend on the position of interaction points in the propagation environment. Thus, since the considered environment is dynamic and the receiver itself is moving, the AoD and AoA are time variant. The direction selectivity is described separately for the transmitter and receiver for all polarization combinations with means of the angular power spectrum (APS). Information about the angular spread and the mean angle can be obtained as well. The APS corresponds with the spatial the inverse Fourier transform of ACF.

## 2.6 MIMO channel

In the previous decade the MIMO technology gained huge popularity and was introduced on the commercial market. This technology takes advantage of multiple antennas at the transmitter (Tx) and receiver (Rx) to increase the system performance by exploiting the spatial dimension [Big07]. The benefits of MIMO are:

- **Array gain** increases the SNR on receiver side as a result of coherent combination of signals.
- **Spatial diversity gain** battles fading by providing the receiver with multiple copies of transmitted signal in space.
- **Spatial multiplexing gain** offers a linear increase in channel capacity by transmitting multiple, independent signals through the channel. If suitable channel conditions (rich scattering) are available and all signals experience the same channel quality, the receiver can separate the signals and the capacity is improved by a multiplicative factor.
- **Interference reduction and avoidance** can be obtained in MIMO by using the spatial dimension to increase the separation between users.

A MIMO channel is described via the channel matrix consisting of the transmission functions of the individual transmit and receive antenna combinations. Assuming a frequency flat channel, the transfer function between the $M$ transmit antennas and the $N$ receive antennas, at a given time instant is represented by the channel matrix $\underline{H}$.

$$\underline{H} = \begin{bmatrix} \underline{h}_{1,1} & \underline{h}_{1,2} & \cdots & \underline{h}_{1,M} \\ \underline{h}_{2,1} & \underline{h}_{2,2} & \cdots & \underline{h}_{2,M} \\ \vdots & \vdots & \vdots & \vdots \\ \underline{h}_{N,1} & \underline{h}_{N,2} & \cdots & \underline{h}_{N,M} \end{bmatrix}. \tag{2.15}$$

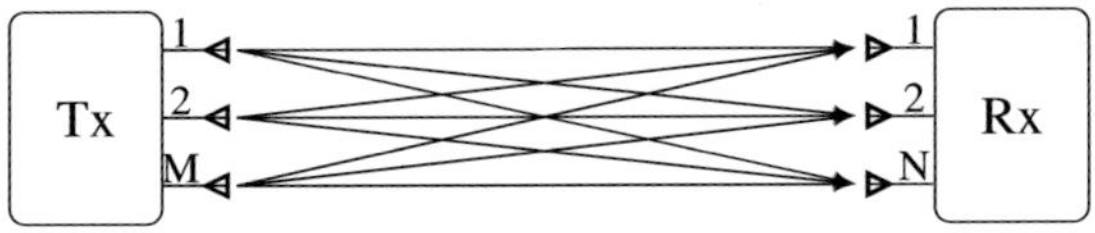


Figure 2.4: Schematic representation of a MIMO communication channel.

Using this notation, the communication channel can be described by:

$$\underline{\vec{y}} = \underline{H}\,\underline{\vec{x}} + \underline{\vec{n}}. \tag{2.16}$$

The transmit signal vector $\underline{\vec{x}} \in \mathbb{C}^{M\times 1}$ and the receive signal vector $\underline{\vec{y}} \in \mathbb{C}^{N\times 1}$ are the inputs and the outputs of the system and $\underline{\vec{n}} \in \mathbb{C}^{N\times 1}$ is the noise signal vector. The noise is additive, zero-mean, complex Gaussian noise. The MIMO communication channel is presented in figure 2.4, the arrows represent $\underline{h}_{n,m}$ elements of the channel matrix.

### 2.6.1 Envelope Correlation Coefficient

A very good and simple metric for a MIMO antenna system's performance is the envelope correlation coefficient (ECC). It is a measure of the correlation between the antennas in a MIMO system. ECC ranges from 0 (no correlation, best MIMO gain) to 1 (identical gain and phase patterns, no MIMO gain). Thus, an antenna system for practical applications should have an ECC lower than 0.5. In its most simple form the ECC value is calculated from antennas' S-parameters based on the formula (4.1) given in [BRC03]:

$$\rho_e = \frac{|S_{11}^* S_{21} + S_{21}^* S_{22}|^2}{(1-(|S_{11}|^2+|S_{21}|^2))(1-(|S_{22}|^2+|S_{12}|^2))}, \tag{2.17}$$

where $S_{11}$ and $S_{22}$ is return loss of antennas A and B respectively, and $S_{21}$ and $S_{12}$ is the isolation between the antenna ports. This value is however not precise and misses information about the antennas' interaction with the channel. The other method to determine antennas' ECC is presented in chapter 5.

### 2.6.2 Channel capacity

Channel capacity is the most important measures used to evaluate the antennas presented in this work. Therefore, it is important to first understand the concept of channel capacity.

In his pioneering work in the late 1940s, Claude Shannon defined basic rules for telecommunication systems. His work refers to data transmission over a channel which adds white noise to the signal. The channel can be a transmission line (cable), a radio channel or an optical transmission path. His theory defines a theoretical upper limit for the data transmission rate over the channel [Big07]. The channel capacity is denoted with $C$ and is defined by the formula:

$$C = \log_2(1 + S/N), \tag{2.18}$$

where $S$ and $N$ are signal and noise power.

The capacity depends only on the SNR at the receiver and is thus a pure channel property, which is independent of the modulation or transmission technique used. As a rule of thumb, if the SNR is doubled (3 dB in logarithmic terms), the capacity is increased by about 1 bps/Hz.

However, in theory it is possible to reach the maximal data rate, in reality it is very difficult to design a channel code reaching the capacity limit. Although great developments in code design have been achieved in the previous decade and data rates very close to the theoretical limit have been demonstrated for very good channels, none of these codes can directly be used for MIMO channels [Big07].

The channel capacity for MIMO is given by formula [Big07]:

$$C = \sum_{i=1}^{N} \log_2\left(1 + \frac{P_{\mathrm{Txi}}}{\sigma^2}\lambda_i\right) = \log_2\left[det\left(I + \frac{SNR}{M} H_{N\mathrm{x}M} H^*_{M\mathrm{x}N}\right)\right], \tag{2.19}$$

where $P_{\mathrm{Txi}}$ is the power at the $i$ transmitter, $\sigma$ is the noise, $\lambda_i$ is the $i$ eigenvalue, $I$ denotes the identity matrix.

## 2.7 Channel modeling

Realistic modeling of multipath propagation is a key component in the antenna evaluation method described in this work. Therefore, a 3D fully polarimetric ray-tracing (RT) tool is utilized to calculate the multipath propagation using the 3D scenario data. The simulator used here was developed at the Institute of Radio Frequency Engineering and Electronics (IHE) of the Karlsruhe Institute of Technology (KIT) [Mau05]. It is a deterministic channel simulator, which means it is based on the physical propagation characteristics of electromagnetic waves in a model of the propagation scenario. This principle is in contrst to stochastic simulators, which try to describe the channel behavior using statistical properties of the power received with a certain delay, doppler shift, angle-of-arrival etc. [Rei13]. Before the RT simulation can be conducted, the environment has to be defined, therefore a brief description of this process is given in section 2.7.1.

### 2.7.1 Environment modeling

In order to get proper results from the RT simulation, the objects (buildings, cars etc.) in the proximity of the transmitter and the receiver should be modeled in detail. These objects create a scenario and can be defined using a 3D map-editor, the detailed 3D information about buildings (including buildings height, roof shape etc.) and streets can be found on the OpenStreetMap (OSM-3D) [OSM].

As can be seen in figure 2.5 a typical building height from 10 to 20 m, sparse vegetation, a dense traffic flow, many parked cars and relatively narrow streets are characteristic for urban environments. In order to obtain the exact results for the interaction between the waves and the environment, the proper values of the EM parameters of the involved buildings, cars and streets have to be set. The map-editor provides functions to set the complex values of the relative dielectric constant $\underline{\varepsilon}_r$ and the relative permeability $\underline{\mu}_r$.

$$\underline{\varepsilon}_r = \varepsilon_r' - j\varepsilon_r'', \tag{2.20}$$

$$\underline{\mu}_r = \mu_r' - j\mu_r'', \tag{2.21}$$

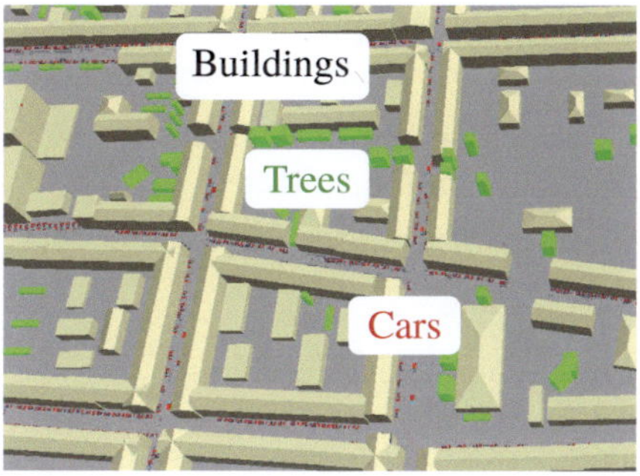


Figure 2.5: An exemplary 3D environment scenario showing a part of downtown Karlsruhe constructed in map-editor.

The imaginary part of the relative dielectric constant includes the information about the dielectric and ohmic losses. The conductivity can be calculated from the dielectric constant as well. The magnetic losses are included in the imaginary part of the relative permeability. Since the structure of a surface has a significant influence on the reflection of an incident wave, the surface roughness is also considered in the RT simulation. The sourface roughness is charaterized with a Gaussian distributed height variation with a mean $\mu$ and a standard deviation $\sigma_h$. In principle the following applies: the rougher the surface, the greater the portion of the energy that is scattered and the smaller the portion that is reflected [Mau05].

In general it is a non trivial problem to model the scattering behavior of the vegetation, due to random distribution of scattering directions. A solution for this was proposed and is explained in [Mau05]. This model is used for trees and bushes, and these objects are represented as homogeneous boxes with defined scattering characteristics.

Since the time-variant behavior of the channel depends not only on the movement of transmitter and receiver, but also on the moving objects in the environment, it is necessary to include the dynamic traffic in the simulation. The individual vehicles' parameters such as speed, acceleration, deceleration and the trajectory can be easily defined in the map-editor. The time progression in the scenario is modeled in discrete time steps, allowing snapshots at the time frames of interest.

### 2.7.2 Wave propagation modeling

The RT tool used in this work is based on geometrical-optics (GO). The basis for this method is the asymptotic behavior of electromagnetic fields at high frequencies (the fields can be described as rays) [AIG90]. This approach is valid as long as the objects in the propagation scenarios are large compared to the wavelength, and the material properties do not change within a wavelength. If these requirements are fulfilled the RT simulator can give precise results within a fraction of time that a finite difference time domain (FDTD) or a finite element method (FEM) solver would require. This is the reason why GO based channel models are also included in commercial V2X tools such as WinProp [Alt] or Remcom Wireless InSite [Rem]. The institute's proprietary RT tool is used in this work, however a commercial tool can also be successfully applied in the presented methodology.

In the RT tool used in this work, the transmitter is considered as a point source for all multipath components. Thus, it is assumed that all observation points are located in the far field of the antenna. The calculation can be divided into two steps. First, an efficient ray search algorithm calculates the course of all multipath components between the transmitter and the receiver. The calculated paths also contain interaction points, at which the rays have interacted with the environment. The reflection paths are traced by using the method of image transmitters (image theory) [Mau05]. In the second step, the transfer factors of the interactions are determined, thus the fully polarimetric channel parameters (amplitude, phase, Doppler etc.) are calculated.

Apart from FSPL, following phenomena are taken into account:

- up to five reflections
- up to three diffraction events
- one scattering event

The combinations of these phenomena are also considered for the resulting paths. The transmission through objects is not considered in the utilized tool. As a result, the RT simulator delivers a position-dependent fully polarimetric information about field strength, phase, time delay and Doppler, as well as AoD and AoA. Each ray represents a multipath component and the combination of

MPCs for given time instant gives a channel impulse response. If a scenario consisting of multiple snapshots is considered, a series of channel impulse responses is obtained.

Although the RT is a deterministic simulator, statistical factors still play an important role. Since the effort to simulate all scenarios occurring in reality would be excessively high, a limited number of scenarios that are as representative as possible is used for the simulation. It is clear that even though a considerate selection is done, the results are not applicable in all environments.

## 2.8 Channel Analysis - Pattern Determination

In order to define radiation patterns prior to the design process, a channel analysis is needed. For the RT simulation 20 different routes based on 3D data of downtown Karlsruhe, each of 150 m length, were chosen. The routes have been divided into two groups based on the morphological structure. The first group is located on the KIT campus (in the center of Karlsruhe) and is characterized by wider streets and more vegetation (see Fig. 2.6a). The second group is placed in a downtown area of Karlsruhe in the vicinity of the campus (see Fig. 2.6b). In this case the traffic is denser and faster. Afterwards, the simulated data is averaged within scenarios from one group. For the simulation omnidirectional point sources are used as the transmit (Tx) and the receive (Rx) antenna. Thus, the Tx and Rx antenna can be replaced with any antenna in the post-processing.

The RT tool provides a data set containing information about amplitude, phase, AoD, AoA etc. of all multipath components (MPCs) for each time step. In order to define optimum radiation patterns, the angular power spectrum (APS) at the vehicular roof-top antenna is extracted from the RT results. Figure 2.7 presents normalized APS for the two groups of investigated scenarios. In both cases a radiation pattern of typical base station (BS) antenna as in [CKC+18], is applied at the Tx side. The Tx power is set to 43 dBm. In case of the campus scenarios, most of the MPCs impinge the Rx antenna from the ± driving direction (see Fig. 2.7a). There are also strong components from right-front and right-rear direction. One has to remember, that if the vehicle would follow the same routes in the opposite direction, the results would mirror to the left side. These results can be explained by the canyoning effect which was already explained

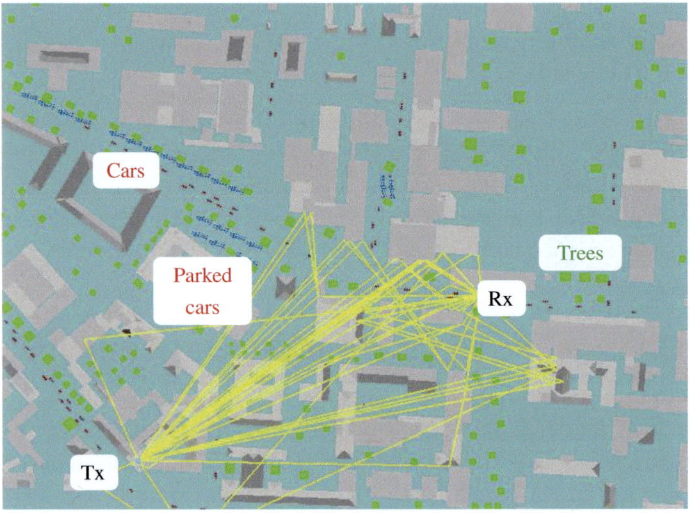


(a) KIT university campus

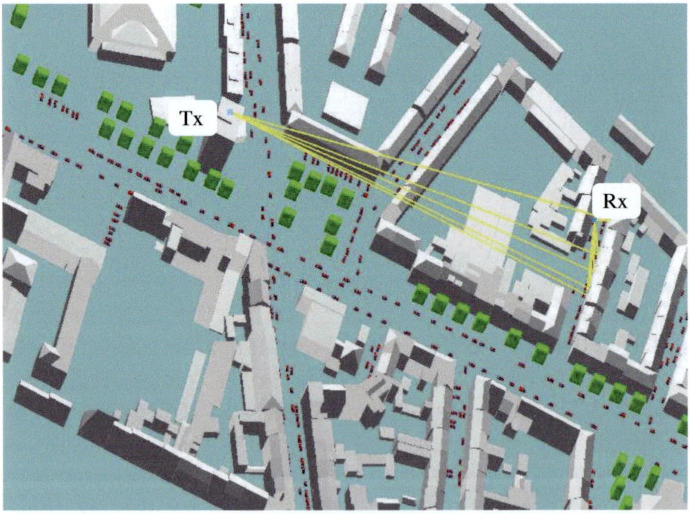


(b) Downtown Karlsruhe

Figure 2.6: Example of two urban scenarios taken into account for the ray-tracer simulation.

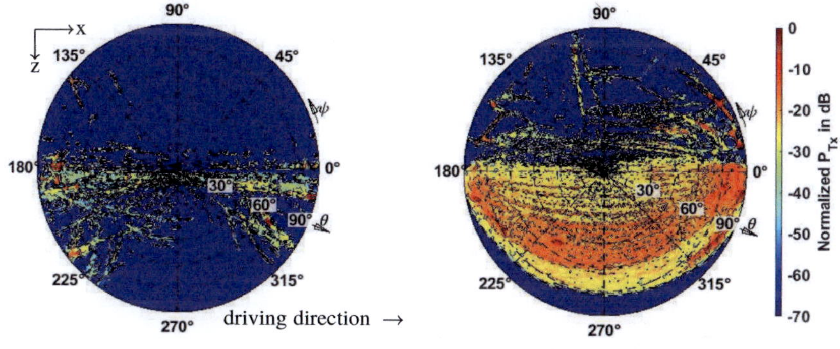


(a) KIT campus (b) Downtown Karlsruhe

Figure 2.7: Simulated APS for a mobile antenna (automotive roof-top antenna). Simulation conducted at 2.49 GHz.

in section 1.3. In the second group of scenarios the strongest MPCs arrive from the angular direction $\psi$ = 170° to 5° in azimuth, and $\theta$ = 45° to 90° in elevation (see Fig. 2.7b). Also in this case the results would mirror along the driving axis to the other side of the vehicle for a drive in the opposite direction. Also in this case the canyoning effect is an explanation for these results. It can be seen that in case of the downtown scenarios there are more MPCs with higher amplitudes. This is mainly due to differences in the morphology of the groups. In the downtown scenarios with narrower streets and less vegetation the MPCs undergo less attenuation. By analyzing these results, it can be assumed that a reconfigurable antenna optimized for automotive applications should switch between beams pointing to the front, rear, right-front, right-rear, left-front and left-rear side of the vehicle. In elevation, the beams should concentrate in an angular range between 15° and 45°.

These results correspond well with general directions resulting from the antenna synthesis presented in [5]. However, we have to remember that the results have some artifacts resulting from statistics (number of scenarios considered). It has to be considered that covering only one direction with one or two antennas, instead of covering more directions simultaneously might give better results. This issue will be investigated in chapter 5.

# 3 Reconfigurable Antennas

Within the second half of the $20^{th}$ century mostly nine different types of antennas were used for wireless applications. These are loop antennas, dipoles/monopoles, slot/horn antennas, microstrip antennas, reflector antennas, dielectric/lens antennas, helical antennas, frequency-independent and log periodic antennas. All of these antennas have some particular properties making them suitable for specific applications. We can say that every other antenna designed in the recent decade is based on the concept of one of these nine categories, adapting the original idea and structure to new applications [Ber07]. However, even if one of these antennas is chosen, its parameters like e.g. polarization are fixed, limiting the system performance. Therefore, more flexible antenna solutions able to react to the changes in wireless channel and battle problems like e.g. fading are required. This is the point at which the concept of reconfigurable antennas was born.

Reconfigurable antennas posses the ability to change their center frequency, impedance bandwidth, polarization and radiation pattern depending on the requirements. It can be done by electrical [ME16, ZNTF18], mechanical [GLG+16, HPD17, 4] or other means e.g. use of liquid-metal [ZGM+18, DGM16]. Nevertheless, the design of reconfigurable antennas constitutes some challenges to the antenna designer. The problem lies very often in retaining the desired performance level in terms of other parameters while reconfiguring one of the parameters. The other issue is feeding the antenna with a direct current (DC) signal, which is necessary for the control of the switching elements. In most of the cases the DC lines influence the performance of the antenna, therefore some counteractive measures have to be taken.

Most of the reconfigurable antennas use p-i-n diodes to switch between the different antenna states [QGWL12, JSJK14, GLW+14]. P-i-n diodes have many advantages like low price, easy control and long lifetime, however they show many vulnerabilities. The major problem is their limited isolation and a relatively high insertion loss. Thus they are mostly used only up to about 10 GHz. However, p-i-n diodes suitable for 5G applications at 28 GHz recently apeared on the market [Mac]. Another problem is their non-linearity. Thanks to the developments in microelectronics within the last decade the micro-electro-machanical system (MEMS) appeared on the market and is another possible solution for state switching [ZMR+13]. MEMS, which are mechanical switches, are linear elements in contrast to p-i-n diodes. Recently the biggest problem of MEMS switches, namely the limited lifetime, was partially solved and off-shelf components for around 10 \$ appeared on the market, thus making it interesting for antenna applications. Another option are the integrated switches. They mostly use complementary metal-oxide-semiconductor (CMOS) elements integrated in reflective or absorptive switches. This solution has a problem of big footprint. Additionally, in most of the cases the MEMS switches as well as the integrated switches are designed for a use in a 50 Ω environment. However, it is mostly not the case if the switch is integrated into the structure of a reconfigurable antenna. The p-i-n diodes seem to be robust against this problem, since they can be considered as variable resistors. Therefore, the p-i-n diodes remain the most popular switching elements used for reconfigurable antennas.

## 3.1 Pattern Reconfigurable Antennas

Among the adjustable parameters of reconfigurable antennas, especially the reconfigurable radiation pattern is very attractive, as it gives a chance to replace expensive and bulky phased arrays. Instead of using multiple radiators with phase shifters, beam steering functionality can be implemented in a single antenna. If we consider that the antenna's radiation pattern is a Fourier transform of the current distribution on the surface of the radiating element or elements, we yield the conclusion that pattern reconfiguration can be obtained by influencing the current distribution. There are several possibilities to do that:

- **shorting pins**, change the current distribution on the radiating element, thus different patterns can be generated [QGWL12] (see Fig. 3.1a),
- **multimode antenna**, in this case different current distributions are obtained by feeding at different points [YZC+16] (see Fig. 3.1b),
- **element switching**, the current distribution of an array is changed by switching between elements [SSA12, GLW+14] (see Fig. 3.1c),

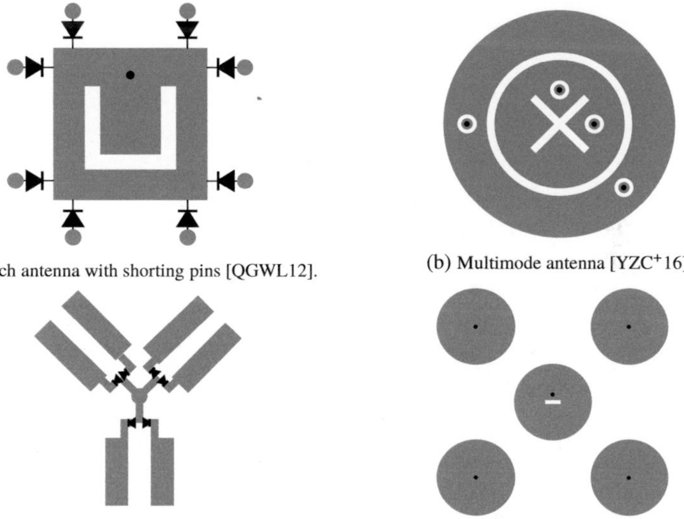

(a) Patch antenna with shorting pins [QGWL12].

(b) Multimode antenna [YZC+16].

(c) Elements switching [GLW+14].

(d) Parasitic elements [JSJK14].

Figure 3.1: Examples of different concepts of reconfigurable antennas.

- **parasitic elements**, in this case the parasitic elements are placed in vicinity of the radiator. The current distribution on the radiator is influenced by that, but more importanttly depending on the impedance of the parasitic element, different current distributions are used to excite the parasitic elments, which itself influence the radiation direction [AA16,JSJK14,ZAH14] (see Fig. 3.1d).

All of these concepts offer reconfigurable patterns and compact design, however the solutions using shorting pins and the multimode antenna are limited in terms of defining the radiation's direction freely. This is because they are based on characteristic modes of the used radiator. In case of element switching and parasitic elements the directions can be defined more freely and thus these concepts are applied and evaluated in this work. Additionally a concept of pattern reconfiguration by switching the current phase between antenna elements is developed and evaluated in this work. Furthermore, the performance of two different switching elements is investigated by means of measurement.

## 3.2 P-i-n Diodes

The p-i-n diode used in this work is a BAP64-02 from NXP. The diodes were measured prior to this work and an equivalent circuit model was established by curve fitting. The s-parameters of the model were thus fitted to the s-parameters obtained in the measurements (the measurement results are presented in section 3.5.1). According to Pozar [Poz12] a series equivalent circuit of p-i-n diode should be used in the forward as well as in the reverse case. Nonetheless, based on the measurement results of and optimization of the switch model in an EM simulator, a series circuit is chosen for the forward case and a parallel circuit for the reverse case. The forward case equivalent circuit consists of a $2\,\Omega$ resistor and a 1 nH inductance in series and the reverse case of a $10\,\mathrm{k}\Omega$ resistor and a 0.13 pF capacity in parallel. These values differ from the values given in the data sheet [NXP], however the parameters evaluated in this work fit the measurement results better.

## 3.3 Reconfigurable Antennas based on Element Switching

In this type of antenna there are multiple radiators which are all connected to the same feeding point (see Fig. 3.1c). The number of reconfigurable patterns that can be realized depends on the number of used elements. The resulting patterns can either be generated by feeding only one element or can result from a combination of fields generated by multiple elements. Depending on the application, such an antenna can either be realized as a planar structure [GLW+14, 23] or as a 3D structure [15]. The function and performance of such antennas will be discussed in more detail in this section (on example of two antennas realized within the scope of this work).

### 3.3.1 A Planar Automotive Antenna based on Element Switching

As already discussed in chapter 1, an automotive antenna for V2X communication should be able to switch between the radiation patterns pointing towards

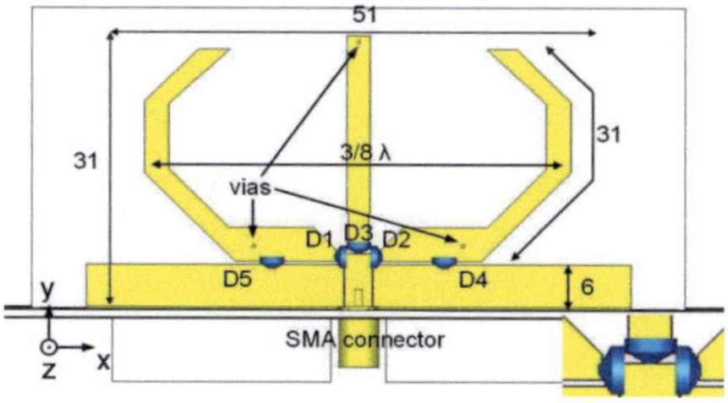


(a) Front side

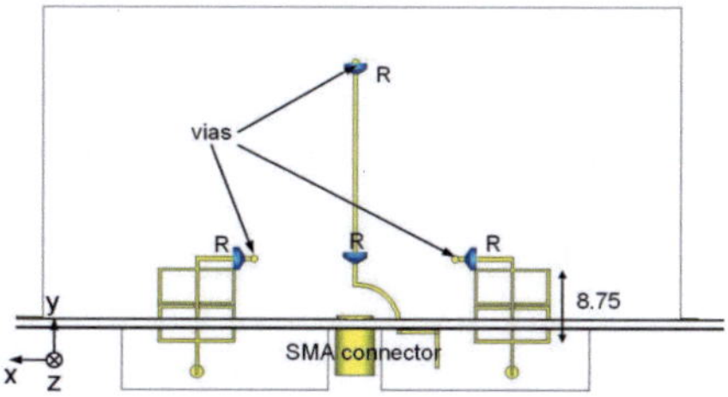


(b) Back side

Figure 3.2: Layout of the planar automotive antenna based on element switching including DC lines, resitors (R) and diodes (D) (dimensions in mm) [23].

front and rear of the vehicle (± driving direction), and towards left and right of the vehicle (to the sides of the car). These patterns can optimize the channel capacity of the system. One way to realize such radition patterns is by using the concept of element switching. A possible solution is proposed by the author in [23].

The antenna consists of three radiators separated by $3/16\lambda_0$ at 2.45 GHz (see Fig. 3.2a). This distance was chosen based on two criteria: first the distance should be kept small for compact antenna construction, second the distance should enable generation of desired patterns and sufficient gain. Thus the distance cannot be too small. The radiators are fed by p-i-n diodes connected to a coplanar waveguide (CPW), which in turn is fed by a an SMA connector mounted through a hole in the ground plane. The middle element is a conventional monopole, while the outer elements have the form of bent metallic lines (see Fig. 3.2a). If the particular diode (D1 or D2) is activated, the center conductor of the CPW feed is connected to the horizontal metallic strip leading to the particular outer element. Thus, the element is fed by a slot line and owing to its bent structure it acts as a tapered slot antenna. Therefore, it exhibits a wide band matching and radiation. However, if the outer elements are used in combination with the middle element they act like monopoles. In the following paragraph the design and functionality of the proposed antenna is discussed.

The antenna is fabricated on a 0.787 mm Rogers RT5880 substrate, which has a relative dielectric permittivity $\varepsilon_r$ of 2.2. The metal thickness is 17 µm. A 6 mm long section of a CPW transmission line is placed in the same plane as the antenna. The signal conductor is 3 mm wide and the gap width is set to 0.1 mm to obtain 50 Ω line impedance. The center conductor is extended by 1.25 mm along the $y$-axis to enable the connection of the three radiators. The edges of the horizontal metallic stripes on the left and right are chamfered in the vicinity of the junction in order to improve the antenna matching (see zoom in Fig. 3.2a). There are three diodes D1, D2 and D3 visible in the figure, each of them is used to activate one desired radiator. Diodes D4 and D5 are placed further to the left and right. As mentioned before, the ground of the CPW section and the horizontal metallic bars leading to the outer radiators establish slot lines used to feed the outer radiators. Even if diodes D1 or D3 are not ON, there is still a portion of the field entering the slot line. By moving the diodes D4 and D5 away from the junction, the short could be transformed to an open at the junction, thus improving the antenna matching. By switching the diodes D4 or D5 ON the desired element can be operated as a reflector. The p-i-n diode used in this work is described in section 3.2.

In order to control the p-i-n diodes DC biasing is needed. The bias network is placed on the back side of the substrate (see Fig. 3.2b) and connected with vias to the front side. As mentioned in chapter 1, DC lines pose some problems for the antennas. Therefore, in this design the following measures are taken to reduce the influence of the lines on E-field. Especially if the lines are placed parallel to the radiated field, a drop in efficiency and changes in radiation pattern might occur. Therefore, the middle line is placed directly behind the central monopole and in the lower part an arc formed to reduce the interference with the electric fields (see Fig. 3.2b). Furthermore, two 5 kΩ resistors are placed in the line to prevent the RF currents from flowing in the line. Considering the outer radiators, the current distribution on the elements was observed first, by means of a 3D EM simulation, and based on the results vias to the back side were placed in regions with the lowest current amplitude.

Table 3.1: Configuration of switching elements for particular reconfigurable states of the planar automotive antenna based on element switching

| | D1 | D2 | D3 | D4 | D5 |
|---|---|---|---|---|---|
| State 1 | ON | OFF | ON | ON | OFF |
| State 2 | OFF | ON | ON | OFF | ON |
| State 3 | ON | ON | OFF | OFF | OFF |
| State 4 | OFF | OFF | ON | OFF | OFF |

If a simple metallic strip leading to a DC pin underneath the ground plane and a resistor as DC choke were used to DC feed the outer elements, it would lead to a decreased antenna performance. Therefore, an alternative solution is required. Using metamaterials for DC biasing can help to avoid these effects. Since metamaterials have frequency-selective properties, they can be used as a filter. In this design a metamaterial structure feeds the DC signal while being itself invisible for electromagnetic waves. The metamaterial cell used is based on electric-LC (ELC) resonators, presented in [SMS06]. The structure is 8.75 mm high and 9 mm wide, and the line width is 0.3 mm. Due to such DC bias design, only a minor reduction of antenna gain by 0.15 dB is observed (instead of around 0.8 dB if simple DC line was used).

The different reconfigurable states can be realized by choosing a particular biasing of the p-i-n diodes. Some attention must be paid to the possible switching between the states, so that only the chosen elements are activated when the voltage is applied and the switching can be realized with minimum number of DC signals. The arrangement and the polarity for all diodes is presented in Fig. 3.2a. If a voltage of −5 V is applied to the DC pin controlling the left element, the diode D1 is in its ON state and the diode D5 is in its OFF state. At the same time the DC signals at the right and the middle element should be both biased with 5 V so that D2 is OFF, D3 is ON and D4 is ON (see Table 3.1). In this case the left and the middle element radiate and the right element acts as a reflector, thus radiation in $-x$ direction (against driving direction) is obtained (see Fig. 3.3a). The maximal gain obtained in this state is 8.31 dBi and the half power beamwidth (HPBW) covers the angular range between $\psi = 115°$ and 245° in azimuth, and $\theta = 42°$ and 78° in elevation. In order to obtain a beam in the driving direction (see Fig. 3.3b), the right element and the middle monopole

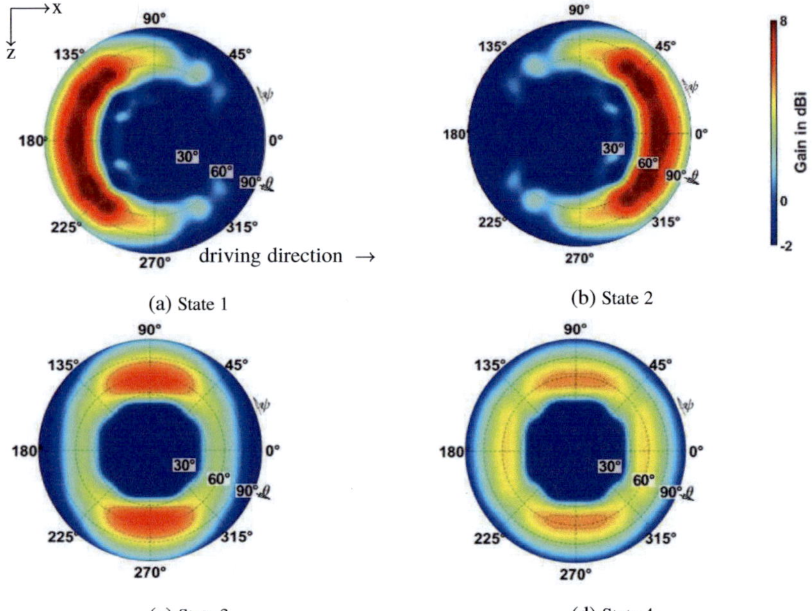


(a) State 1

(b) State 2

(c) State 3

(d) State 4

Figure 3.3: Simulated radiation patterns of the planar automotive antenna based on element switching at 2.45 GHz [23].

have to be connected to signal line of the CPW, while the left element is shorted. The obtained gain has the same level as for state 1 and the HPBW is between $\psi = 295°$ and $65°$ in azimuth, and between $\theta = 42°$ and $78°$ in elevation. In state 3 the main radiation lobes are pointing to the sides of the car ($\pm z$ direction). In this case both outer elements are active while the middle element is not fed. The maximal gain for this state is 6.5 dBi and the HPBW in azimuth for the two beams is between $\psi = 22°$ and $168°$, and between $\psi = 202°$ and $338°$. Finally, in state 4, if only the middle element is fed and the outer elements are neither active nor in reflector mode, a nearly omnidirectional pattern with a maximum gain of 5.3 dBi is realized (see Fig. 3.3d).

After evaluating the antenna's performance a prototype of this antenna was fabricated. A photolithographic process is used to form the antenna structure on the surface of the Rogers 5880 substrate. The p-i-n diodes (BAP64-02 from

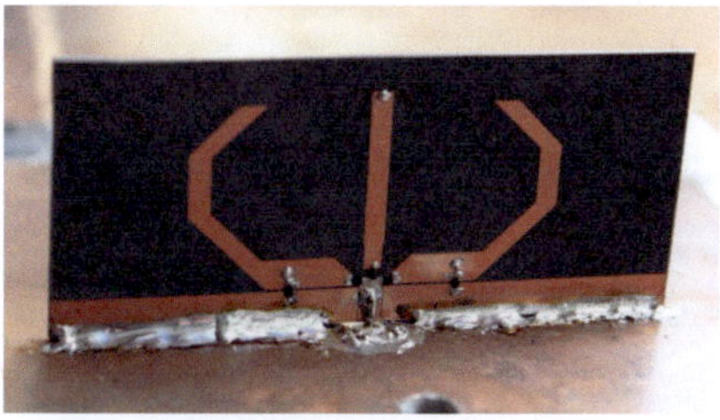

(a) Antenna prototype

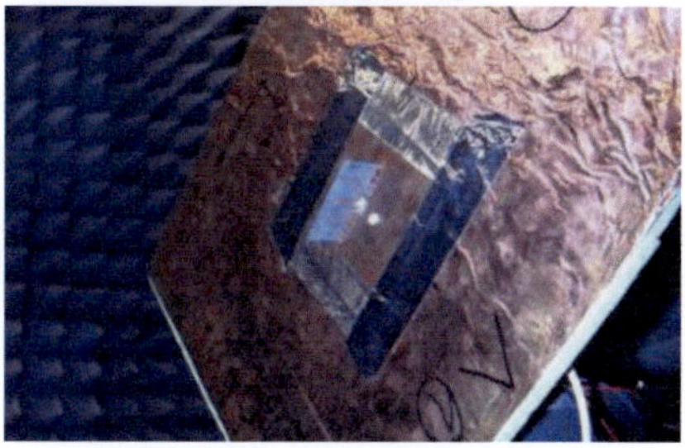

(b) Antenna on the big ground plane

Figure 3.4: Fabricated prototype of the planar automotive antenna based on element switching and the setup used for measurement in the unechoic chamber.

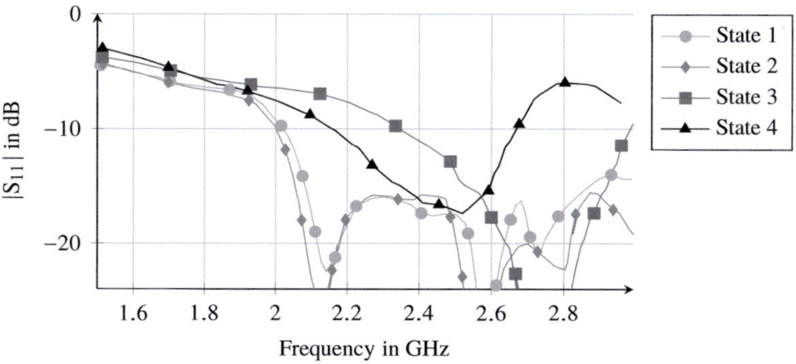


Figure 3.5: Measured S-parameters of the fabricated prototype for the four different reconfigurable states.

NXP) and resistors are soldered manually. The PCB board with the antenna is eventually mounted on a ground plane, which consists of a metal plated FR-4 board with three milled slots. These slots are necessary to mount a SMA connector and insert the PCB parts with DC feeding lines through the ground plane. Thus all the ports are placed underneath the ground plane. A bigger ground plane ($500 \times 600\,\mathrm{mm}^2$) is used to mimic the car roof in the measurements (see Fig. 3.4).

First the return loss of the antenna was measured. A return loss better than 6 dB is obtained in the frequency range from 1.88 to 2.8 GHz (see Fig. 3.5). The states 1 and 2 show good matching in a wider frequency range, which is due to the tapered slot construction. Furthermore, it can be noticed that for state 4, where only the central monopole is active, the obtained matching is worse than for the other states. Furthermore, the frequency is slightly lower than for states 3 where both outer elements are active.

The next step in the evaluation of the antenna performance was the measurement of gain and radiation pattern in the institute's anechoic chamber. Measured directivity patterns correspond very well with the simulation. The direction of measured beams and their HPBW are similar to the simulated ones. For state 1 a beam in $-x$ direction (towards rear of the vehicle) can be observed (see Fig. 3.6a), and the measured realized gain is 8.4 dBi and exceeds the simulated value by 0.1 dB. The results for the state 2 are also very similar to the simulation (see Fig. 3.6b). The gain is the same as for state 1. Considering states 3 and 4, the measured gain is respectively 6.5 dBi and 5.9 dBi and the form of the patterns resembles the simulated ones. Remarkably, the measured gain is higher than the simulated one. It can be explained by having a look at the gain over frequency. There is a ripple with a magnitude of of about 1 dB on this curve. Thus the realized measured gain can differ by 0.5 dB from the measured value. Considering the measurement at 2.45 GHz this value could be deduced since a maximum of ripple appears at this frequency. This ripple can be explained as a combination of antenna matching and poor output matching of an amplifier used for antenna measurement (transmitting antenna). Thus a standing wave between antenna and amplifier is generated.

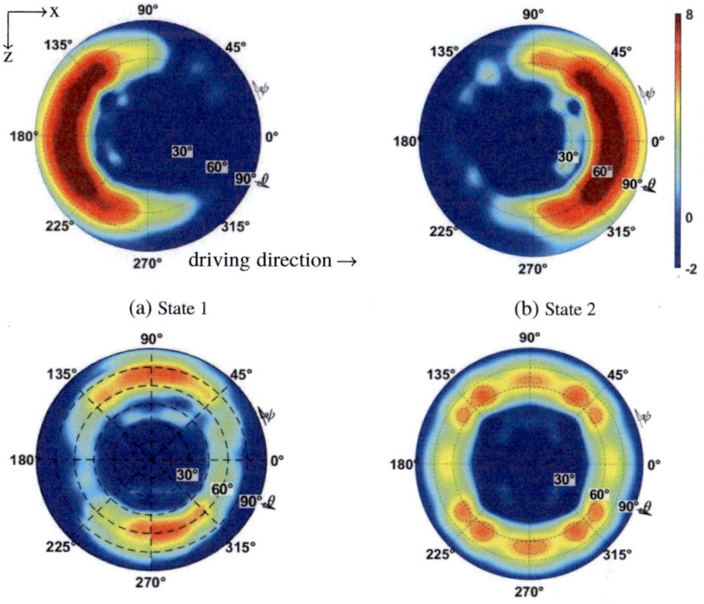


(a) State 1

(b) State 2

(c) State 3

(d) State 4

Figure 3.6: Measured radiation patterns of the fabricated prototype of the planar automotive antenna based on element switching at 2.45 GHz.

## 3.3.2 A 3D Automotive Antenna based on Element Switching

Another possibility to realize an antenna that covers the directions crucial for V2X communication is to use a 3D structure instead of planar one. In [15] the author proposed an antenna consisting of two planar antenna parts, which are orthogonal to each other. In this case the pattern reconfiguration is as well realized by switching between elements. This antenna generates four different beams which are shifted by 90° in azimuth in respect to each other.

Each of the antenna parts consists of two radiating elements (see Fig. 3.7a and 3.7b) and a CPW transmission line is used for feeding (similar as in section 3.3.1). The radiating elements consist of a metallic patch and an inverted-

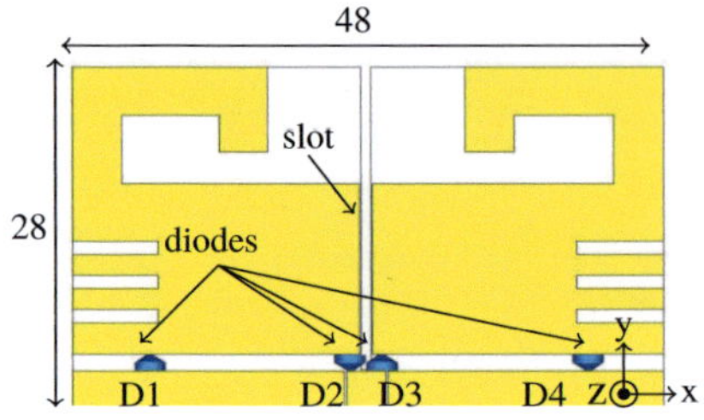


(a) Antenna part 1

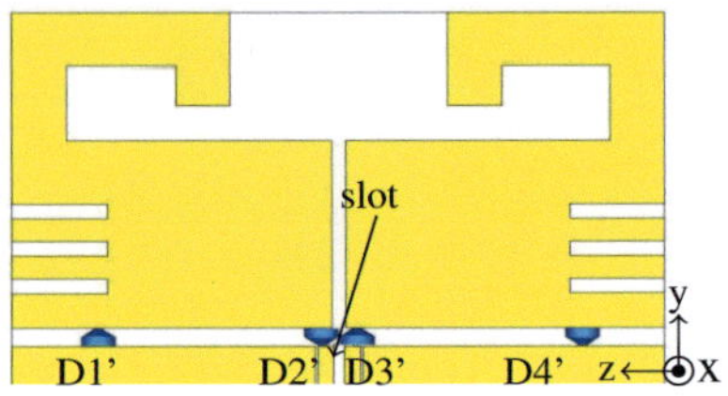


(b) Antenna part 2

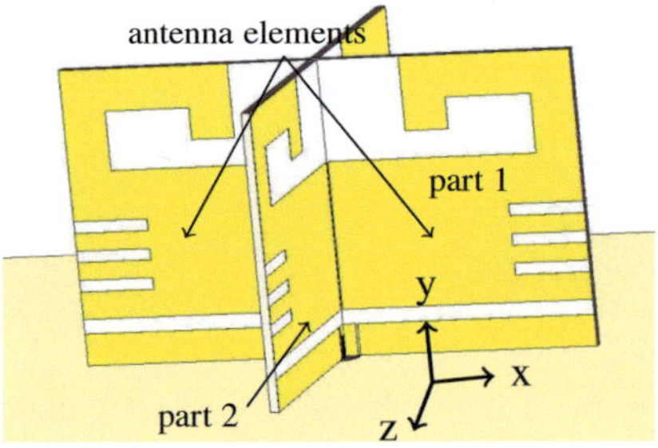


(c) Assembled antenna

Figure 3.7: Layout of antenna parts and assembled model of the 3D automotive antenna based on element switching (dimensions in mm).

L structure on top. The task of this inverted-L structure is to decrease the antenna's dimension. The distance between the elements is 1 mm (about 0.01 $\lambda_0$) and thus they are tightly coupled, what is important for the function of the non-fed element as a reflector. This distance was optimized based on the results presented in [HSN08]. The corrugations placed on the outer edge of the elements improve the antenna matching and increase the front-to-back ratio (FBR) by improving the current distribution on the reflector. The gap between the CPW line and the radiating elements is set to 1.4 mm and it was optimized for low return loss. There is a slot in the substrate of each antenna part one from the top (see Fig. 3.7a) and one from the bottom (see Fig. 3.7b). Due to these slots, both antenna parts can be slid into one another so that the parts are orthogonal (see Fig. 3.7c). The size of the assembled antenna is $48 \times 28 \times 48$ mm$^3$.

Table 3.2: Configuration of the switching elements for the four reconfigurable states of the 3D automotive antenna based on element switching

| | D1 | D2 | D3 | D4 | D1' | D2' | D3' | D4' |
|---|---|---|---|---|---|---|---|---|
| State 1 | OFF | ON | OFF | ON | OFF | OFF | OFF | OFF |
| State 2 | ON | OFF | ON | OFF | OFF | OFF | OFF | OFF |
| State 3 | OFF | OFF | OFF | OFF | OFF | ON | OFF | ON |
| State 4 | OFF | OFF | OFF | OFF | ON | OFF | ON | OFF |

To switch between the antenna elements, two opposite polarized p-i-n diodes per radiating element were used. One connects the antenna element to the signal line of the CPW line, the other one to the ground. As for the antenna shown in scetion 3.3.1, the BAP64-02 diodes were used. State 1 is realized if the diode D2 is switched ON and diodes D1 as well as D3 are switched OFF, and diode D4 is ON (all of the diodes of the orthogonal antenna part are biased in their OFF stete). In this case the left element (see Fig. 3.7a) is fed and radiates the electromagnetic field, while the right one acts as a reflector. Due to the strong coupling between both elements, current opposite in phase to the current on the radiator is generated on the reflector. The reflector element is connected to ground with diode D4 in order to enable proper current distribution on the element. Due to the connection to the ground the currents on this element are reduced. This contributes to lower radiation to the back and increased efficiency. In this state the antenna radiates in $-x$ direction (against driving direction) (see Fig. 3.8a). The maximum gain obtained in this state is 6 dBi and the HPBW covers the angular range between $\psi = 100°$ and $260°$ in azimuth, and $\theta = 20°$ and $85°$ in elevation. A beam in the driving direction (see Fig. 3.8b) is realized in state 2. Also in this case the maximum gain is equal to 6 dBi. For states 3 and 4 beams directed to the sides of the vehicle ($+z$ and $-z$ direction respectively) are generated (see Fig. 3.8c and 3.8d). The maximum gain is likewise 6 dBi. All of the p-i-n diode configurations are given in table 3.2. It can be observed that this antenna exhibits a lower gain and a broader beam if compared to the states 1 and 2 of the antenna discussed in section 3.3.1.

Since the diode pairs at each antenna element are oppositely polarized with respect to one another (see Fig. 3.7a) a simple DC feeding of the designed antenna is possible. A coaxial bias-T is included in the RF signal path. Both, the

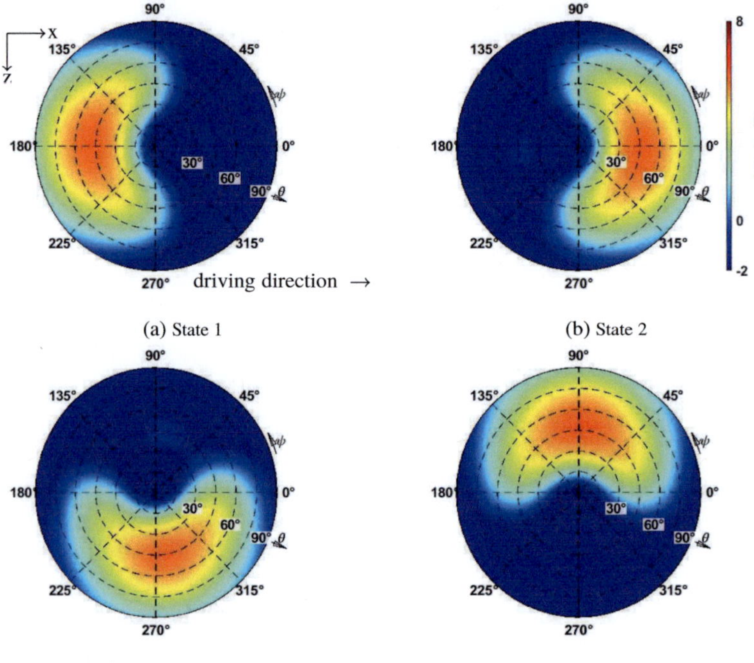


(a) State 1

(b) State 2

(c) State 3

(d) State 4

Figure 3.8: Simulated radiation patterns of the 3D automotive antenna based on element switching at 2.6 GHz.

signal line and the outer conductor are grounded for the DC signal. By applying a positive or a negative DC signal to the antenna elements the particular diodes are activated. At the same time the orthogonal antenna part should be connected to the ground for the DC signal and thus the diodes are switched OFF. The model of the p-i-n diode used for the simulation is the same as presented in section 3.2.

In the next step a prototype of the discussed antenna was fabricated. The fabrication process as well as the used substrate (0.787 mm thick Rogers RT5880) is the same as for the design from section 3.3.1. In this case the assembled antenna was mounted on a brass plate used as a ground plane (see Fig. 3.11). After soldering the assembled antenna to the ground plane, the DC lines consisting

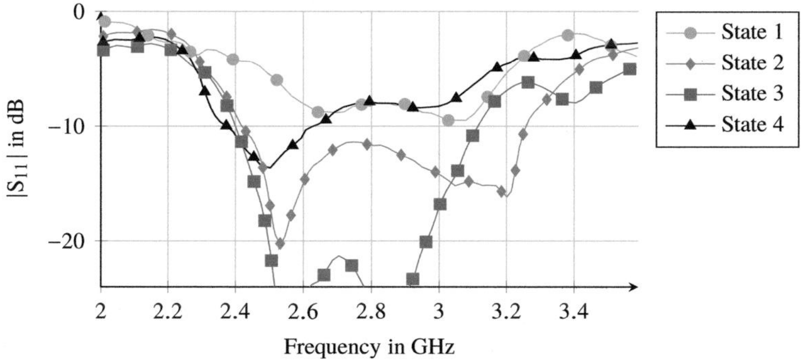


Figure 3.9: Measured S-parameters of the fabricated prototype of the 3D automotive antenna based on element switching for the four reconfigurable states.

of a 5 kΩ resistor and a wire (guided through a hole in the ground plane) are soldered to the corners of antenna elements. Depending on element either −5 V or 5 V voltage is applied at the DC lines.

The measured bandwidth of the antenna covers the frequency range from 2.5 to 3.12 GHz (6 dB bandwidth). A return loss better than 6 dB within a wider band is obtained for states 2 and 3. The states 1 and 4 show worse results in terms of return loss, especially state 1 (see Fig. 3.9). Looking at the good results for the other two states, it can be assumed that this is a result of fabrication inaccuracies. To enable mechanical connection of the antenna, a vertical slot has to be cut in the signal line of antenna part 2. In order to connect the split parts of the signal line a wire is lead through a hole (0.6 mm diameter) in the substrate and the CPW line of antenna part 1, thus making the structure asymmetric. This influence the states realized by both antenna parts and can explain the differences between the four states.

Afterwards the antenna was measured in the anechoic chamber. Two important elevation cuts ($\psi$=0° and $\psi$=90° in azimuth) were measured (see Fig. 3.10). In the plots the directive patterns and FBR of around 5 dB is well visible. The maximum measured gain at 2.6 GHz is around 4 dBi. This gain is nonetheless similar to the simulations results in which circuit models were used for the p-i-n diodes. The gain for state 1 is lower than for the other states, what is expected

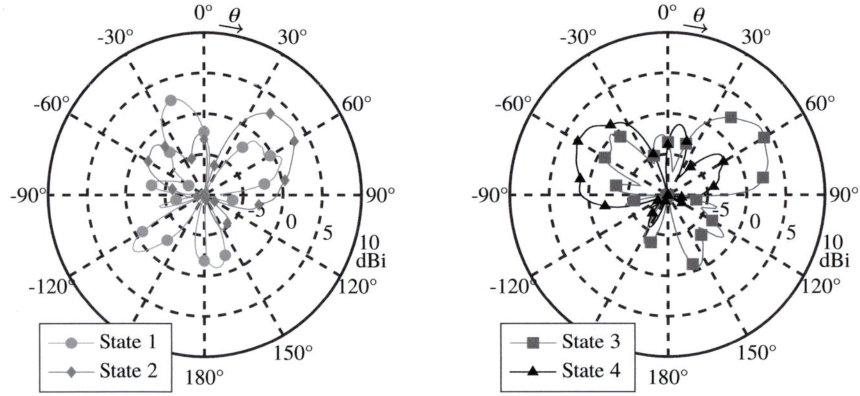


(a) $\psi$=0°, $xy$-plane (b) $\psi$=90°, $yz$-plane

Figure 3.10: Two measured elevation cuts of the radiation pattern of fabricated prototype for four reconfigurable states of the 3D automotive antenna based on element switching at 2.6 GHz.

Figure 3.11: Fabricated prototype of the antenna.

after evaluation of the return loss measurements. The HPBW is similar as in the simulation and covers angular range of $\theta = 35°$ to $85°$ in elevation.

## 3.4 Reconfigurable Antennas Utilizing Parasitic Elements

This type of antennas normally consists of a primary radiator placed in the middle and parasitic elements placed around it. The parasitic elements should be loaded with a variable impedance so that they can switch between reflector and director mode, similar as in case of a Yagi antenna. These parasitic loads can either be realized by loading the elemens with varactors or simply connecting them to the ground using p-i-n diodes (switching between short and open). Such an antenna is known as an electronically steerable passive array radiator (ESPAR) [KO05, ZAH14]. Depending on the number of parasitic elements a different number of beams can be realized, thus this solution offers high level of flexibility. The author proposed a simple planar solution utilizing parasitic elements in [18].

### 3.4.1 A Wideband Antenna Utilizing Parasitic Elements

The antenna presented in this section resembles a CPW fed slot monopole antenna [PK10]. The central element is a round wideband monopole (see Fig. 3.12), adjacent to it are two semicircular metallic patches which are each divided in

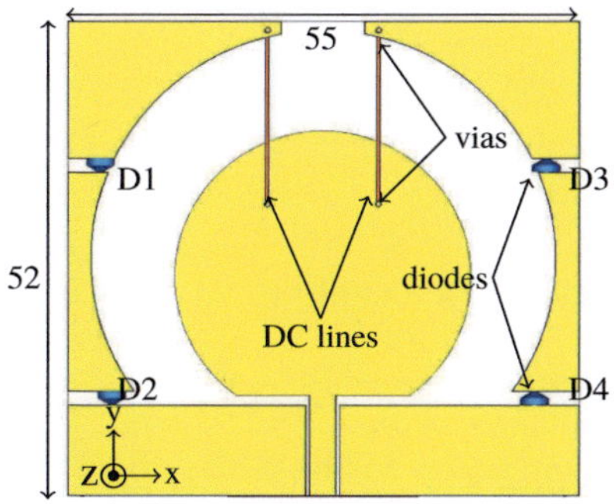


Figure 3.12: Layout of the antenna model utilizing parasitic elements. DC lines are placed on the back side of the substrate (dimensions in mm) [18]. Front side metallization is given with yellow color and back side metallization with orange.

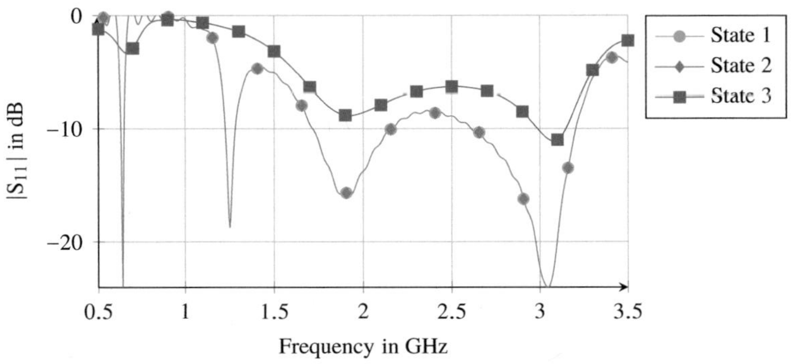


Figure 3.13: Simulated S-parameters of the presented antenna utilizing parasitic elements for three reconfigurable states.

two parts. Normally these metallic patches are connected to the ground and to each other establishing a wideband slot, however in this case they function as parasitic elements. This is the reason why they are separated by slots. In order to control the function of parasitic elements p-i-n diodes are placed in these slots connecting the parts to each other and to the ground (see Fig. 3.12). Furthermore, there are two 0.1 mm wide DC lines placed on the back side of the substrate which are connecting the monopole and the parasitic elements using vias. The lines are placed in a section of the circular slot, where the E-field is weak, and thus their influence on antenna performance is minimized. A further measure to prevent the field from interaction with the DC lines is including an inductor acting as RF choke in each of them (similar as in section 3.3.1). Due to such construction, the antenna can be easily fed with a DC signal using a coaxial bias-T placed in the RF signal line. The evaluation using an EM simulator showed that the influence of these lines on the antenna's performance is negligible. The overall antenna size is $55 \times 52\,\text{mm}^2$ ($0.46\,\lambda_0$ by $0.43\,\lambda_0$ at 2.45 GHz).

As already mentioned, there are two p-i-n diodes per parasitic element. Once again diodes as presented in section 3.2 are utilized. The diodes of one element are in the opposite polarization to the diodes of the other element. In this configuration state 1 is active if a negative voltage is applied at the bias-T, and diodes D1 and D2 are OFF while diodes D3 and D4 are ON. In this case the right parasitic element acts as a reflector while the lower part of the left one acts as a director, the beam is directed in $-x$ direction (see Fig. 3.14). The maximum

Table 3.3: Configuration of switching elements for three reconfigurable states of the presented antenna utilizing parasitic elements

| | D1 | D2 | D3 | D4 |
|---|---|---|---|---|
| State 1 | OFF | OFF | ON | ON |
| State 2 | ON | ON | OFF | OFF |
| State 3 | OFF | OFF | OFF | OFF |

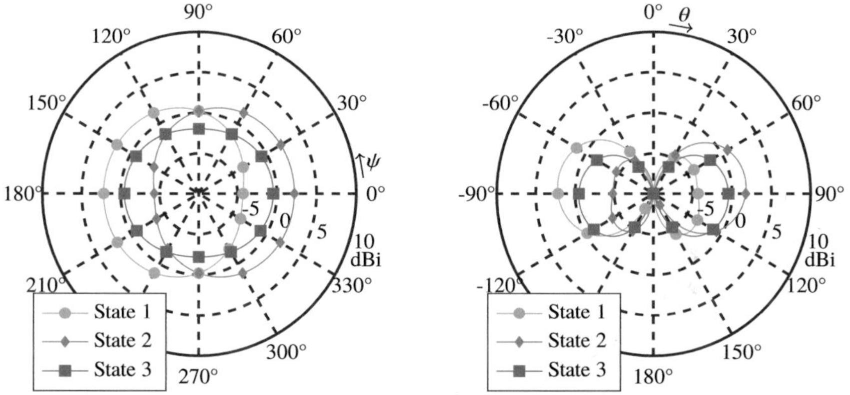


(a) Azimuth cut for $\theta$=90°, $xz$-plane

(b) Elevation cut for $\psi$=0°, $xy$-plane

Figure 3.14: Simulated radiation pattern of the presented antenna utilizing parasitic elements for three reconfigurable states at 2.45 GHz.

gain is about 3 dBi and the FBR is 8 dB. The HPBW covers the angular range between $\psi = 92°$ and 268° in azimuth, and $\theta = -50°$ and $-120°$ in elevation. In contrast, if a positive voltage is applied at the bias-T, diodes D1 and D2 are ON, and D3 and D4 are OFF. In state 2 the left parasitic element functions now as a reflector and the left one as a director, the beam is directed in +$x$ direction (see Fig. 3.14). Due to the symmetry of the structure, the same gain and FBR are obtained and the HPBW covers the angular range between $\psi = 272°$ and 88° in azimuth, and $\theta = 50°$ and 120° in elevation. State 3 is activated if no voltage is applied at the bias-T and all diodes are OFF. In this state the gain slightly exceeds 0 dBi. The diode configuration for the different states is given in table 3.3. If

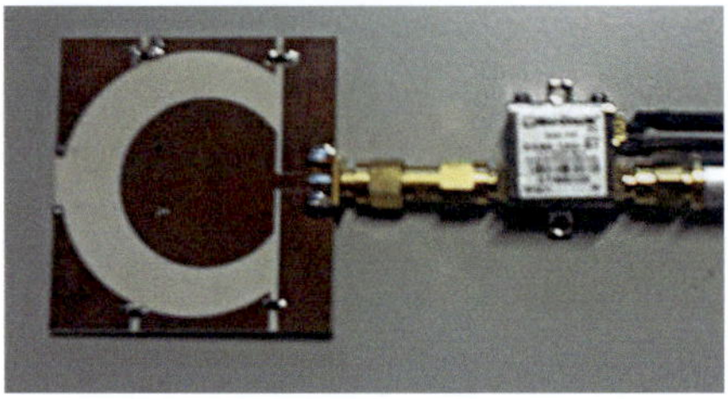

Figure 3.15: Fabricated prototype of the antenna utilizing parasitic elements including a coaxial bias-T (Mini-Circuits ZX85-12G-S+).

6 dB return loss is used for evaluation of the antenna's bandwidth, the frequency range from 1.5 GHz to 3.2 GHz is covered for all states (see Fig. 3.13).

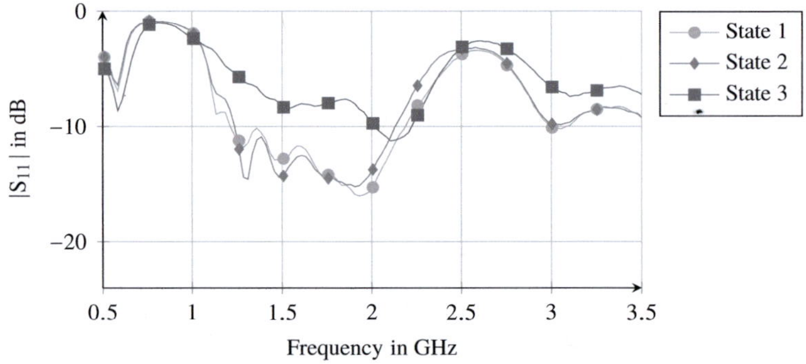


Figure 3.16: Measured S-parameters of the presented antenna utilizing parasitic elements for three reconfigurable states.

After evaluating the antenna with the aid of an EM simulator, a prototype of the antenna was fabricated. In this case a 0.8 mm thick Rogers 4003 ceramic substrate was used ($\varepsilon_r$ = 3.5). After etching the antenna structure, the SMA connector, the p-i-n diodes and the vias were soldered manually (see Fig. 3.15). The return loss of the prototype was then measured and although the results show some variance compared to simulation the matching between 1.4 GHz and 2.45 GHz is achieved for all states (the calibration of network analyzer was conducted including the bias-T). In general, the return loss curve is moved towards lower frequencies (see Fig. 3.16). Next the gain and directivity patterns

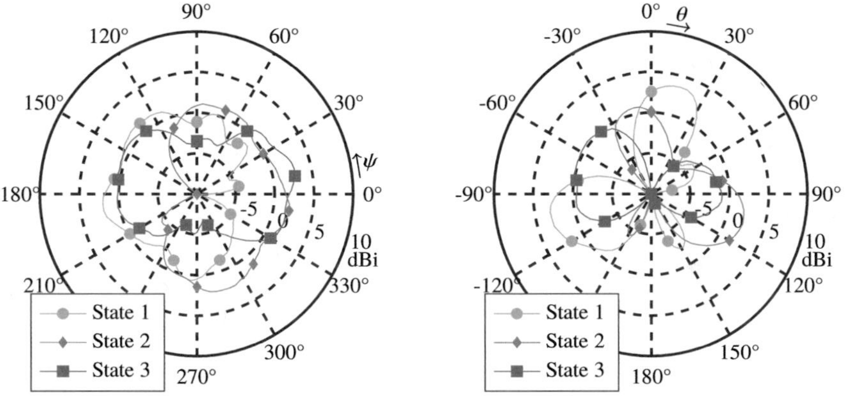


(a) Azimuth cut for $\theta$=90°, $xz$-plane

(b) Elevation cut for $\psi$=0°, $xy$-plane

Figure 3.17: Measured radiation pattern of the presented antenna utilizing parasitic elements for three reconfigurable states at 2.45 GHz.

were measured. The maximum gain achieved for state 1 is 2.5 dBi and a similar value is achieved for state 2 (see Fig. 3.17). The azimuth cuts of radiation pattern for these states are very similar to the simulated ones, except for a minimum at the direction opposite to the direction of the main beam (see Fig. 3.17a). The HPBW for states 1 and 2 is 125° and 180° respectively. More differences are observed for the elevation cuts (see Fig. 3.17b). Instead of a wide beam in the desired direction, there is a narrow beam (45° HPBW) in this direction and an additional beam in the angular range $\theta$ = 0° and ±30°. This differences in form of radiation pattern might be caused by the measurement setup and coupling with the coaxial cable which was not considered in simulation. The gain for state 3 is around 1 dBi and exceeds the simulated value by 1 dB. In this case the measured pattern corresponds well with the simulation.

## 3.5 Reconfigurable Antennas based on Phase Switching

Apart from the four main methods to realize reconfigurable radiation patterns there is another possibility that was investigated and used in the scope of this work. Pattern steering and switching can be realized by using multiple radiators and switching the phase difference between these radiators. The operating principle resembles this of a phased array, however in this case no additional expensive and complicated phase sifters and variable gain amplifiers are needed. The phase shifting can be realized by using varactors or switching between different transmission lines. However, in this case the phase shifting elements are implemented in the antenna structure.

### 3.5.1 Pattern Reconfigurable Antenna Utilizing Multiple Monopoles

An example design using this principle was presented by the author in [19]. Two monopoles separated by about $\lambda_0/2$ (free space wavelength at 2.45 GHz) are the crucial parts of this antenna. The left monopole is wider and fed with an offset feed for better matching (see Fig. 3.18a). This monopole is loaded on the top with a line connected to the second monopole. This line itself divides into two lines: lower and upper one, which both lead to the right monopole, however one line is connected to the top of this monopole and the second one to the bottom. If the right monopole is fed with the lower of the lines, this monopole will be fed with a signal that is in phase with the signal applied at the left monopole (0° phase difference). To obtain this signal a line with length $n\lambda$ is needed, where $n$ is an integer and $\lambda$ is the wavelength in the substrate. The obvious choice is $\lambda$ and thus since the left monopole is $\frac{1}{4}\lambda$ the line feeding the second monopole has to be $\frac{3}{4}\lambda$ long. In order to fit the line into the given space, the line has a sinus form. The sinus form has an additional function, simulation results prove there is only little radiation from the line thanks to the this form. The same length is needed for the upper line in order to feed the monopole with a signal in opposite phase (180° difference). The phase difference occurs since the monopole is fed from the top. The switches enabling the antenna's reconfiguration are placed in the right part of the two lines. The distance to the junction is around $\lambda/2$, therefore if one switch is open, an open at this branch is visible for the wave at the junction and the reflection is minimized. The overall antenna size is $66 \times 27\,\text{mm}^2$.

In order to activate state 1, the upper switch S1 is closed (ON) and the lower switch S2 is open (OFF), then the monopoles are fed with signals in opposite phase. In this case, independent of the distance between the radiators, a destructive interference along the $z$-axis occurs. Furthermore, since the distance

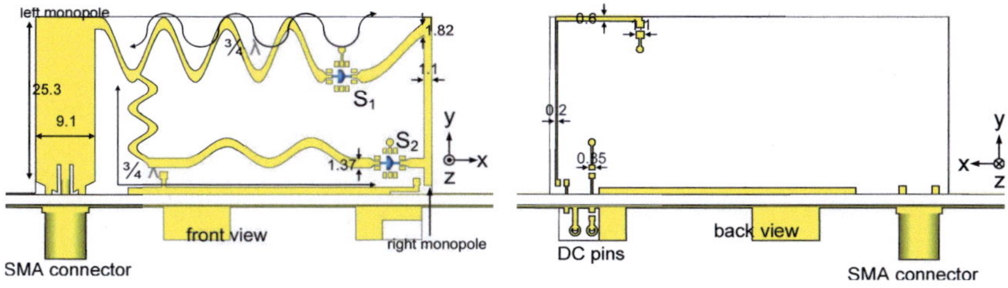


(a) Front side

(b) Back side

Figure 3.18: Layout of the antenna utilizing multiple monopoles including DC lines and diodes (dimensions in mm) [19].

between the two monopoles is $\lambda_0/2$ the fields interfere constructively along the $x$-axis (see Fig. 3.19a). The maximum gain of 6.9 dBi is obtained in $-x$ direction, and is 0.6 dB higher than in $+x$ direction. This happens, since the monopoles are fed in series and the left one is the first one in the series and gets more energy from the fed signal. The HPBW covers angular ranges $\psi = 307°$ to $53°$ and $120°$ to $240°$ in azimuth, and $\theta = 56°$ to $83°$ and $52°$ to $81°$ in elevation for directions $+x$ and $-x$ respectively. State 2 is activated when the upper switch S1 is open (OFF) and the lower switch S2 is closed (ON), then the monopoles are fed with signals in phase. In this case, a constructive interference occurs along the $z$-axis (see Fig. 3.19b), and a destructive interference along the $x$-axis (due to distance between antennas). The obtained gain is 7.6 dBi and is the same for both beams. The HPBW covers angular ranges from $\psi = 67°$ to $170°$ and from $221°$ to $296°$ for directions $-z$ and $+z$ respectively in azimuth, and $\theta = 46°$ to $78°$ in elevation. An asymmetry occurs against the $z$-axis, the beam is wider in $-x$ direction. The explanation is the same as for state 1.

Two electrical switches were considered for this design: p-i-n diodes and MEMS. First, two test-boards one for MEMS (Radant RMSW200HP-QFN) and one for p-i-n diode (BAP64-02) were fabricated and measurements were conducted to determine proper values for the circuit models (see Fig. 3.20). This investigation proved that for the used configuration there are only minor differences in isolation and insertion loss between p-i-n diode and MEMS. Although it would be expected, that MEMS offers better results than p-i-n diode, since no connection to the ground pad of the MEMS is available on the antenna, the results

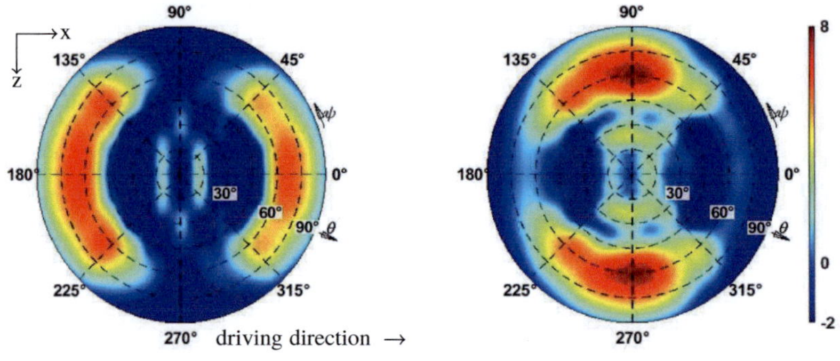


(a) State 1

(b) State 2

Figure 3.19: Simulated radiation patterns of the presented antenna utilizing multiple monopoles at 2.45 GHz.

deteriorate and get close to those of diode. Therefore there is no advantage of using more expensive MEMS switches.

DC feeding of MEMS switches is a bit more complex than feeding p-i-n diodes. First, they need to be supplied with a −90 V DC signal. Second, due to their

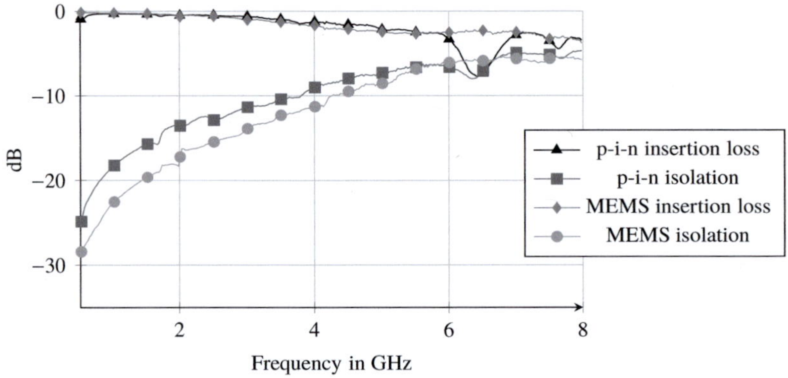


Figure 3.20: Measured parameters of the BAP64-02 p-i-n diode (biased with forward current of 10 mA and reverse voltage of −1 V) and Radant RMSW200HP-QFN MEMS.

construction and QFN package the DC feeding is done using separate DC pins unlike for diodes. Therefore, additional DC lines connecting the pins beneath the ground plane (see Fig. 3.18b) to both of the MEMS switches are placed on the back side of the antenna. A vertical line placed behind the right monopole is used to fed the upper MEMS. Just next to it is a short line used to fed the lower MEMS. Such an arrangement of the lines and resistors implemented into the lines minimize the influence of the DC lines on the antenna properties. These DC lines are removed from the layout for the antenna with p-i-n diodes. Furthermore the solder pads for the p-i-n diodes differ from those for MEMS, however the rest of the layout remains unchanged.

Simulation results of a model using MEMS show that antenna is matched between 2.37 GHz and 2.67 GHz for both states, if 6 dB return loss is considered (see Fig.3.21a). For model using p-i-n diodes matching is obtained between 2.36 GHz and 2.62 GHz (see Fig.3.21b). In both cases the matching for state 2 is better, most probably due to less reflection in the lower feeding line. There are no big differences observable in terms of matching between the model with MEMS and the one with diodes.

In the next step two prototypes, one using MEMS and one using p-i-n diodes, were fabricated. The structure was etched on 0.787 mm thick Rogers RT5880 substrate. The p-i-n diodes (BAP64-02) were soldered manually and the MEMS were soldered by dispensing solder paste on the pads and reflow soldered in an oven. The antenna was then mounted perpendicular to a metal plated FR-4 board with a size of $150 \times 100\,\text{mm}^2$ using two slots milled in the board (see Fig. 3.22). This structure is then placed on a big ground plane $500 \times 600\,\text{mm}^2$ for the measurements. The prototype using MEMS is fed with DC signal using DC pins placed underneath the ground plane. A self-constructed DC converter and control board are used to supply the prototype with $-90\,\text{V}$. In case of the prototype with p-i-n diodes a standard bias-T connected to a laboratory power supply was used to feed diodes with the DC signal.

A shift in frequency compared to the simulation results is visible in the measurement results of the input reflection (see Fig. 3.21). However, the shift is less than 50 MHz and 100 MHz for diode and MEMS respectively. Despite the shift in frequency, good matching between the measurement ans the simulation is achieved for both states. Finally, measurements in an anechoic chamber were conducted and realized gain and directivity patterns were evaluated. The maximum gain for state 1 at azimuth $\psi = 0°$ is about 4.5 dBi (compared to simulated

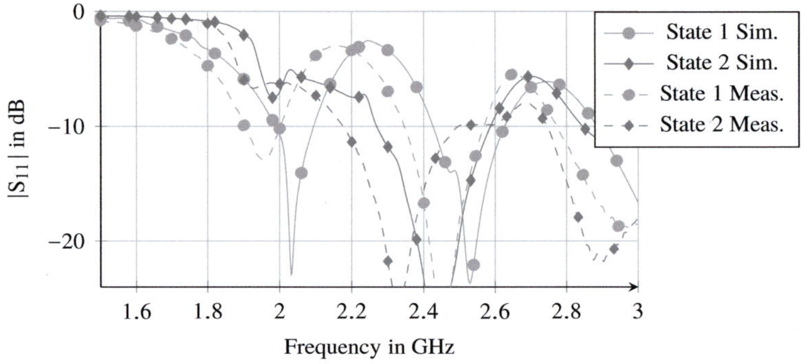


(a) Antenna using MEMS

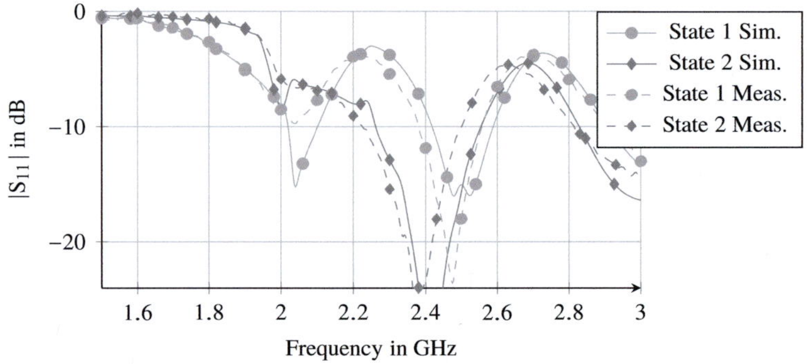


(b) Antenna using p-i-n diodes

Figure 3.21: Simulated and measured S-parameters of the presented antenna utilizing multiple monopoles for two reconfigurable states.

Figure 3.22: Fabricated prototype of the presented antenna utilizing multiple monopoles utilizing MEMS [19].

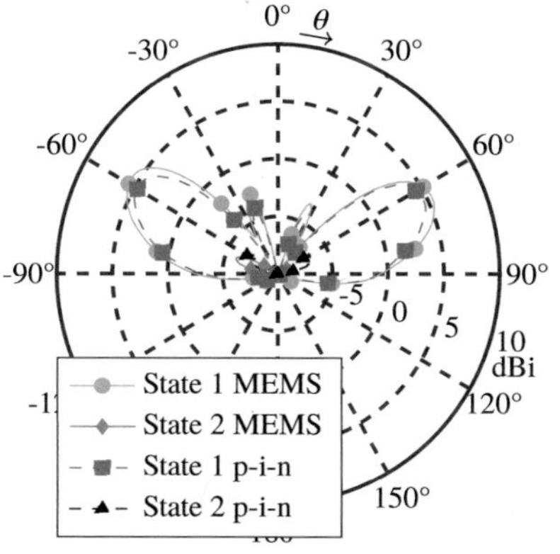


(a) Elevation cut for $\psi$=0°, $xy$-plane

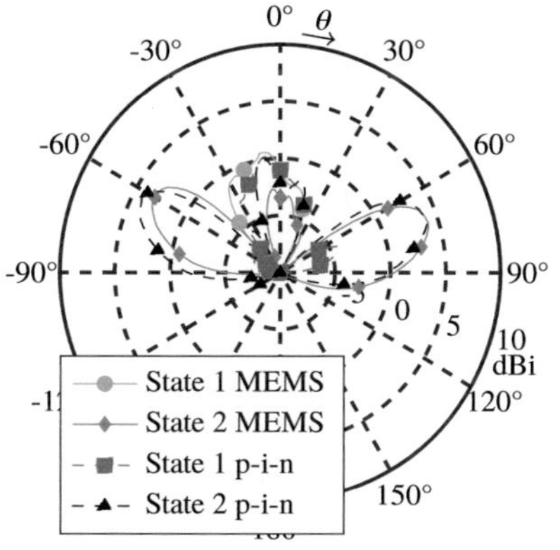


(b) Elevation cut for $\psi$=90°, $yz$-plane

Figure 3.23: Measured radiation pattern of two prototypes of the presented antenna utilizing multiple monopoles one using MEMS switches and one using p-i-n diodes for two reconfigurable states at 2.45 GHz.

6.4 dBi) and it is almost equal for the prototype using MEMS and p-i-n diode (see Fig. 3.23a). The gain for state 2 in this elevation cut is lower than −4 dBi. Considering the elevation cut at $\psi$ = 90° for state 2 a maximum gain of 5.2 dBi and 4.9 dBi is obtained for the prototype using MEMS and p-i-n diodes respectively, compared to the simulated 7.2 dBi (see Fig. 3.23b). The gain for state 1 is in this case lower than −7 dBi. Comparing the measured values with the simulation a difference of about 2 dB is visible in all cases. The difference for the prototype using p-i-n diodes is slightly higher than for the one using MEMS, this is due to higher losses than expected in simulation. Additionally, it is possible that the p-i-n diodes model is not accurate enough and a different phase shift between the radiators, compared to the simulation occurs, thus causing a shift of the radiation beams in azimuth. Another reason for differences between simulation and measurement is the frequency shift in the matching. The measurement inaccuracies and switch losses can be accounted for further losses. Nevertheless, all the measured curves correspond well in their form with the simulated ones.

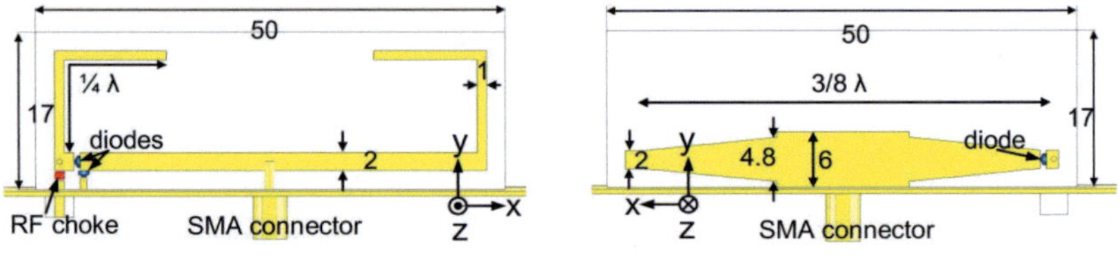


(a) Front side

(b) Back side

Figure 3.24: Layout of the antenna utilizing balun for beam switching including DC lines and diodes (dimensions in mm) [16].

### 3.5.2 Pattern Reconfigurable Antenna Utilizing Balun for Beam Switching

In the previous subsection (3.5.1) switching between two phase delay lines was proposed as a solution for generating different phase shifts between the antennas. However, a difference in gain between two beams in one state and a shift in the beams' direction for the other state was observed due to feeding in series. An alternative is to feed the radiators in parallel. Thus the unequal amplitude distribution can be omitted. In this case the radiators should be placed symmetrically around the feeding point. However, if the antenna should stay compact and the desired distance between the radiators is set to around $\lambda_0/2$, placing delay lines with different lengths between feed point and radiator appears problematic. A solution to this problem was proposed by the author in [16] and is described in this section.

In the discussed antenna two inverted-L radiators are used instead of monopoles to reduce the antenna height (see Fig. 3.24). The distance between the radiators is reduced to $\frac{3}{8}\lambda_0$ compared to the previous design, thus the antenna's size can be further reduced. The reduction of antenna separation does not impair the beam switching, since the generated phase shift in state 1 can be tuned to new design parameters and there is only a negligibly decrease of maximum gain, due to slightly stronger coupling between the radiators. The overall antenna size is $50 \times 17\,\mathrm{mm}^2$ and the size is reduced by about 50 % making it a more attractive candidate for implementation in the shark-fin housing.

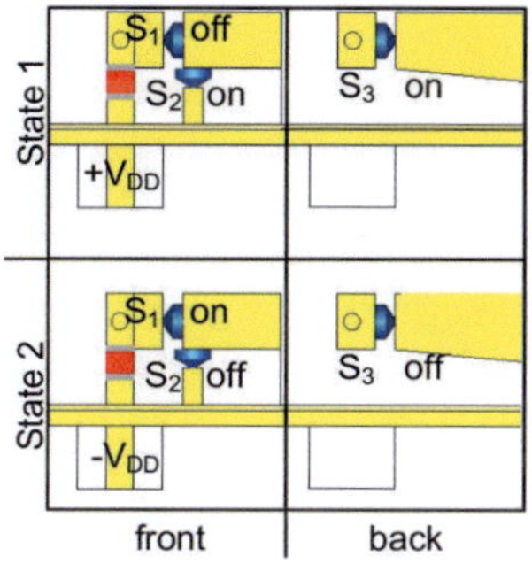


Figure 3.25: Switch configuration on the front and back side of the substrate for the presented antenna utilizing balun for beam switching [16]. The p-i-n diodes are denoted with S and the red square symbolizes a RF choke.

The most important difference compared to the antenna from section 3.5.1 is the feeding network. The radiators are fed in parallel using a horizontal microstrip line connected in the middle with a coaxial line (see Fig. 3.24a). This solution saves space since an additional T-junction is omitted. A crucial issue to obtain two switchable patterns as defined before is the possibility to generate a phase difference of either 0° or 180° between the radiators. In this case the use of a linear tapered microstrip balun in connection with p-i-n diodes is proposed to generate the phase shift between the elements (see Fig. 3.24b). This type of baluns is very often used to generate a symmetric signal at the input of ultra wide-band antennas [SUE12, XYN09]. This type of balun works in the following way: a microstrip transmission line (which is unbalanced) is placed over a ground plane which is gradually tapered along propagation axis until it is the same shape as the metal strip that lies above it. At the end of the transition, the two lines are identical in shape, and thus a balanced microstrip line is obtained. Because of the available space the balun was reduced in width compared with conventional designs known from literature ($0.06\lambda_0$ compared to $0.2\lambda_0$ width in [SUE12]). In the next step it was optimized to generate an optimal phase shift between the two inverted-L antennas. Proper phase difference is crucial for the generation of the pattern in the ± driving direction (along the $x$-axis). To conduct the optimization, the antennas were replaced with discreet ports and the phase difference between these ports was evaluated. As a result either 0° or 180° phase shift was generated at the left end of the line (compared to the right end)

Table 3.4: Configuration of switching elements of the presented antenna utilizing balun for beam switching for two reconfigurable states.

| | S1 | S2 | S3 | $V_{DD}$ | $V_{\text{bias-T}}$ |
|---|---|---|---|---|---|
| State 1 | OFF | ON | ON | 10 V | 1 V |
| State 2 | ON | OFF | OFF | −10 V | 0 V |

depending on the state of p-i-n diodes (see Fig. 3.25). To improve the matching in state one, a stub with a p-i-n connecting the front side line to the ground was placed. Whenever the switch S1 is OFF (switch S3 is ON), the switch S2 is ON shorting the line to the ground to improve the return loss. Metallic pads on each side of the substrate (see Fig. 3.25) are connected with a via in order to enable switching between the states. The pads are connected with counter-wise polarised p-i-n diodes to the microstrip line at the front and the tapered line at the back side.

The line used for the DC supply is placed in the lower left corner of the structure (see Fig. 3.24a) in order to minimize the influence of the DC lines on the antenna properties. The line is decoupled from the RF signal with a 10 kΩ resistor which is used as a choke. Due to its high impedance it acts as an open for the RF signal. At the same time it has only negligible influence on the DC supply because the current floating through the p-i-n diodes is relatively low (1 mA). The pin used for DC supply is placed under the ground plane where it is shielded from the EM fields. For proper function of the antenna a voltage is also fed by a SMA connector with aid of a coaxial bias-T. The control voltage configuration for both states is presented in table 3.4.

In order to activate state 1 (beams towards front and rear of the vehicle) the radiators have to be fed with a signal in opposite phase (180° phase difference) the switch S1 has to be OFF, while switches S2 and S3 are both ON (see Fig. 3.25). Thus, the left antenna is fed by the tapered line at the backside of the structure, and the current on the left antenna is opposite in phase to the current on the right element. As a result, a constructive interference along the $x$-axis and destructive interference along the $z$-axis occurs (see Fig. 3.26a). The maximum gain of 6.8 dBi is obtained in the $+x$ direction, and is 0.7 dB higher than in the opposite direction. This happens, because the left radiator is fed by the tapered line. Since the currents in the tapered line are not as strong as those

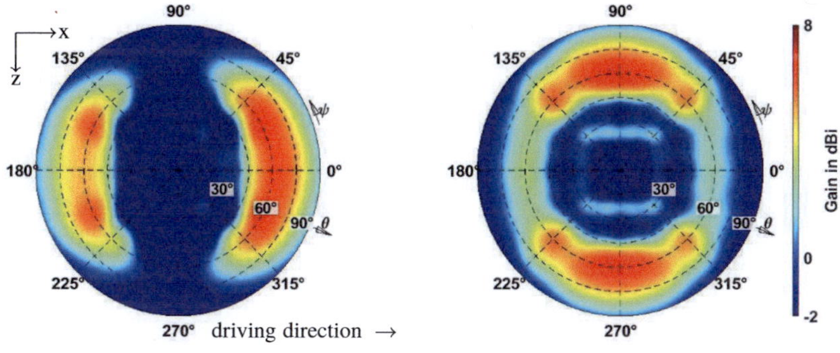


(a) State 1

(b) State 2

Figure 3.26: Simulated radiation patterns of the presented antenna utilizing balun for beam switching at 2.45 GHz.

in the microstrip line and additional losses due to the diode appear (compared to the right radiator), the signal fed to the left radiator has a lower amplitude. Therefore, this problem was not solved compared to design from section 3.5.1. The HPBW covers angular ranges from $\psi = 302°$ to $58°$ and from $130°$ to $230°$ in azimuth, and $\theta = 47°$ to $82°$ and $44°$ to $84°$ in elevation for directions $+x$ and $-x$ respectively. In case of state 2 the switch S1 has to be ON and the switches S2 and S3 are both OFF (see Fig. 3.25). The radiating elements are now fed with signals that are in phase, and the two radiators interfere constructively along the $z$-axis. The maximal gain is 7.1 dBi and the difference between the beams is only 0.1 dB. The HPBW covers the angular ranges $\psi = 36°$ to $144°$ and $216°$ to $324°$ in azimuth, and $\theta = 48°$ to $78°$ in elevation. The shift in the beams' direction is eliminated compared to design from section 3.5.1. The HPBW increased though which is due to smaller distance between radiators.

Simulation results show that the antenna is matched between 2.37 GHz and 2.63 GHz for both states, if 6 dB return loss is considered (see Fig. 3.27). Thus, a slightly smaller bandwidth than for the antenna from section 3.5.1 is obtained.

After stimulative evaluation of the antenna's performance a prototype was fabricated. The antenna was etched on Rogers RT5880 substrate with thickness of 0.787 mm. The p-i-n diodes (BAP64-02) and the resistor were soldered manually to it. The antenna was then vertically mounted through two slots in metal

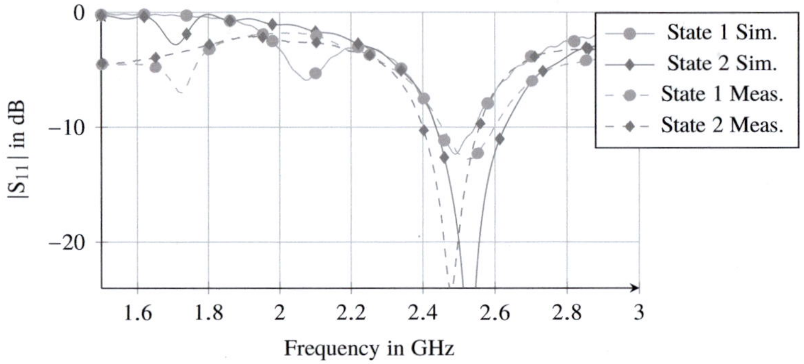


Figure 3.27: Simulated and measured S-parameters of the presented antenna utilizing balun for beam switching for two reconfigurable states.

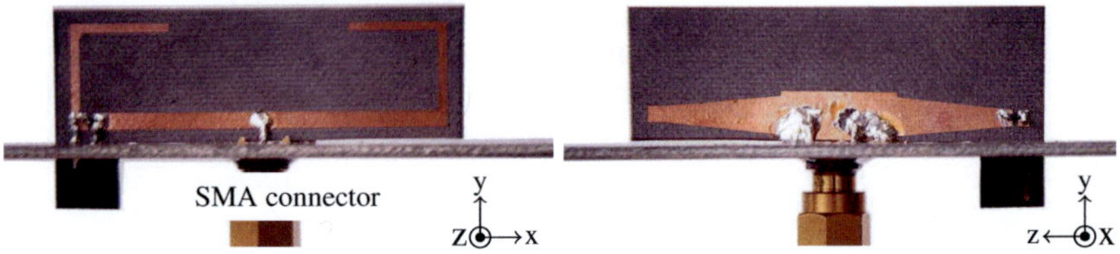


(a) Front side

(b) Back side

Figure 3.28: Fabricated prototype of the presented antenna utilizing balun for beam switching.

plated FR-4 board with a size of $150 \times 100\,\text{mm}^2$ which is used as a ground plane (see Fig. 3.28). This structure can easily be attached to a bigger ground plane for sake of measurements in anechoic chamber.

The curves of the measured S-parameters correspond well with the simulation, however the center frequency is slightly shifted by about 50 MHz towards lower frequencies for state 2 (see Fig. 3.27). Similar tendency was observed for the antenna presented in section 3.5.1. Nevertheless, it it is still around 2.45 GHz. In next step the prototype was measured in an anechoic chamber. The maximum gain measured for the state 1 in the $+x$ direction is about 5 dBi and the gain in the opposite direction is lower by about 2 dB (see Fig. 3.29a). This trend was also the case for the simulation results, however the difference was about

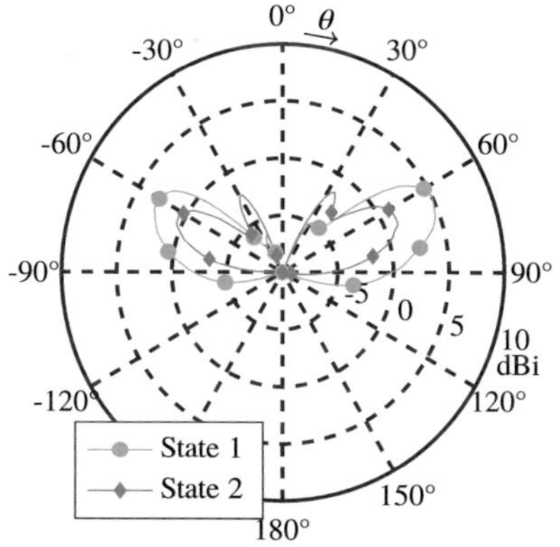


(a) Elevation cut for $\psi$=0°, $xy$-plane

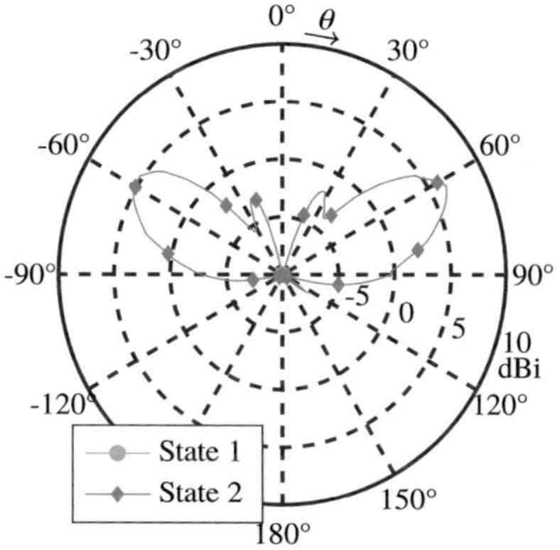


(b) Elevation cut for $\psi$=90°, $yz$-plane

Figure 3.29: Measured radiation pattern of the prototype of the presented antenna utilizing balun for beam switching for two reconfigurable states at 2.45 GHz.

0.7 dB. In this plane (elevation cut at $\psi$= 0°) maximal gain for state 2 is about 1 dBi. If the elevation cut at $\psi$= 90° is considered, the maximum gain for state 2 is 6.5 dBi (in $-z$ direction) and about 1 dB lower in the opposite direction. For state 1 the gain in this plane is lower than −10 dBi. A degradation of about 1.5 dB between the measurement and the simulation can be observed for both states. This difference is caused mainly by the additional phase shift caused by the diode and inaccuracies in the fabrication.

## 3.6 Conclusion

In this chapter different approaches for realization of pattern reconfigurable antennas were presented. The generated radiation patterns are optimized for automotive urban propagation scenarios. Among different concepts three approaches are studied: element switching, parasitic elements and phase switching. These concepts are explained in detail on example designs. Depending on the concept, different number of reconfigurable states can be generated and different patterns are realized.

To generate patterns covering the ± driving direction simultaneously or beams pointing to both sides of the vehicle at the same time, a method based on phase switching seems to be most attractive. Considering the two designs, based on this concept, presented in this chapter, the antenna from section 3.5.2 has a smaller dimension and higher measured maximum gain (it is robuster against fabrication inaccuracies). Therefore, this concept will be further developed and incorporated in a MIMO antenna in the next chapter.

If the desired directions (front, rear, left, right etc.) should be covered only one at the time, the element switching method gives very good results. A 3D solution presented in this section was susceptible to the fabrication process and required a big volume. Thus, only a planar solution will be further investigated in the next chapter.

The concept of using parasitic elements has a promising potential for more flexibility if the antenna size is increased in the 3$^{rd}$ dimension and further parasitic elements are introduced (compared to the design from section 3.4). This concept will be further explored in the next chapter. The cost of increasing the flexibility is the increased volume of the antenna. This issue will be reviewed as well in the next chapter.

Furthermore, a comparison between MEMS and p-i-n diodes was investigated on an example of one of the designs. The evaluation shows that in practical applications there is no significant difference in performance between p-i-n diodes and MEMS. Thus, the higher effort connected with the use of MEMS does not bring any improvement.

# 4 Pattern Reconfigurable MIMO Antennas

The reconfigurable antennas presented in chapter 3 bring a potential improvement in performance of the automotive communication systems. However, as defined in modern communication standards like 4G [ETSa] and 5G [ETSb] SISO (single-input single-output) systems should be replaced by MIMO systems (the advantages of MIMO were discussed in chapter 1). In order to make reconfigurable antennas compatible with these standards and provide even further increase in performance, these antennas should be incorporated in MIMO systems. While establishing a multiple antenna system the engineer is faced with some challenges like e.g. keeping the coupling between the antennas low while reducing the overall antenna size or designing decorrelated patterns for the antennas.

This chapter presents different designs of pattern reconfigurable antennas suited for MIMO applications. The antenna systems are designed based on the experience collected during development of antennas presented in chapter 3. Furthermore, different decoupling structures and their influence on the antennas' performance are investigated in detail.

## 4.1 Multiple Antenna System using Back-to-Back Balun for Pattern Switching

The first system discussed in this chapter was proposed by the author [14]. The requirements on reconfigurable patterns for this antenna system are the same as in chapter 3 (beams covering front and rear, and sides of the vehicle). The single antenna consist of two radiators separated by $0.5\lambda_0$ (at 2.6 GHz) fed in parallel and the reconfigurable patterns are generated by switching the phase between the radiators (similar as the design from section 3.5.2). However, in this case a

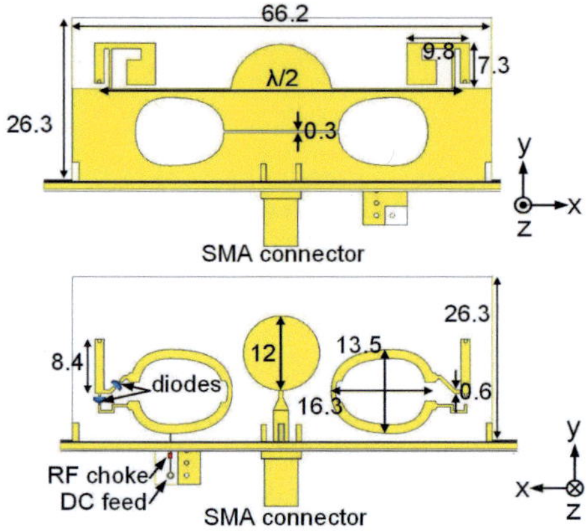


Figure 4.1: Layout of a single antenna using back-to-back balun including DC lines and diodes (dimensions in mm) [14].

different feeding network is used. Details of the single radiator design, MIMO configuration and decoupling structure are explained in this section.

### 4.1.1 Single Element

A starting point of the antenna design is in this case the feeding network. It has to ensure a proper phase difference between the radiators and to enable the switching. A linear tapered balun was successfully applied by the author in [16]. Beside the advantage of compact size, such a design has a drawback of generating a phase difference of either 0° or 180° only. This is due to the construction of the balun where the output is a parallel stripline. Thus the phase difference between the lines is 180°. A more flexible solution able to generate any phase difference and thus any desired directivity pattern is needed. A solution for this problem is to use a back-to-back balun [Gup96]. In most of the cases a back-to-back balun is used to convert the wave from microstrip-to-slotline or vice-versa. However, if a circular slot is used for slotline-to-microstrip transition (see Fig. 4.1), one can get a symmetrical signal at the output lines (0° and 180° phase). The symmetric signal at the output appears due to currents circulating along the edges of the round slot. These currents couple to the microstrip line encircling the slot on

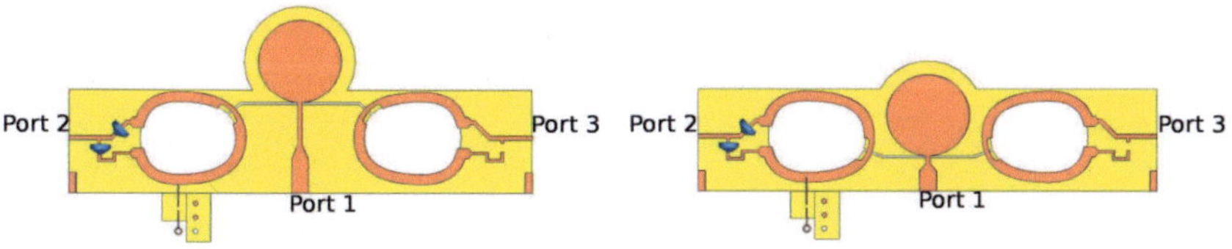


(a) Structure 1 (b) Structure 2

Figure 4.2: Modification of feeding structure for various phase differences between the ports [14].

the other side of the substrate, thus generating the currents in this line. The coupling depends on the substrate thickness, and the width of the microstrip and its alignment with the slot edge. It is also important that the circumference of the round slot should be around $\frac{1}{2}\lambda_0$ at the design frequency in order to ensure matching at the input port and an efficient transition between the microstrip and the slot mode. The structure presented in this work is fed in the middle with a coaxial line connected to the vertical microstrip line (see Fig. 4.2). This microstrip line (in turn) feeds the horizontal slotline establishing a T-junction. In this case a circular stub is placed at the end of the microstrip (open to short transition, as current maximum in the plane of slot enables efficient coupling to the slotline). The impedance of the microstrip line and the slotline at the transition point are tuned to have the same values to ensure efficient coupling. An additional advantage of the microstrip-to-slotline transition is decoupling of the DC signal from the RF port. Thus no further DC blocks are required, and as a result no further losses are included in the system. Initially circular slots were used, however they are transformed to elliptic ones in order to reduce the height of the feeding network (see Fig. 4.1 and Fig. 4.2). Further modification can be seen at the point where the slotline is connected to the elliptic slot. In this case the width of the microstrip line was reduced for better matching. The microstrip (orange in Fig. 4.2) is located partially above the metallized surface (yellow in Fig. 4.2), the other part is placed above the elliptic slot. It is important to tune the width of the line placed above the elliptic slot to ensure a good transition of the signal.

To obtain a phase difference of 0° and 180° the slotline should be placed in the middle of the height of the elliptic slot (see Fig. 4.1). Arbitrary phase differences between the output ports can be obtained after some modification. If the point at which the horizontal slotline coming from the T-junction is connected to

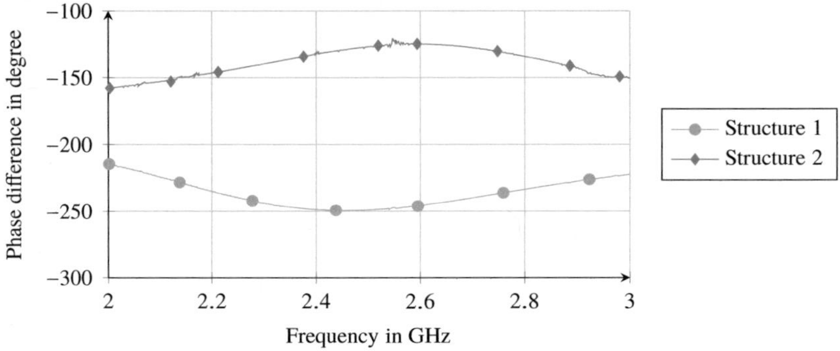


Figure 4.3: Phase differences between the ports of the structures from Fig. 4.2.

the elliptical slot changes (see Fig. 4.2), the phase difference can be modified. Simultaneously, the position of the microstrip line leading to the output ports remains unchanged (case 1). Another option is to change the position of the microstrip lines leading to the ports, while leaving the slotline unchanged (case 2) or combining both for even more design freedom (case 3).

Since, the mechanism of the modification is not changing between the three cases, and only a different phase variation is obtained, only the influence of the horizontal slotline position, as presented in Fig. 4.2, is explained in more detail. The modification leads to a change in length of the upper and lower path of microstrip, which is coupled with the circular slot, and thus the phase difference at the outputs of the microstrip line changes as well. For the case study the upper path is hardwired to port 3 and at port 2 the paths can be switched using p-i-n diodes. However, whenever the upper path is chosen at port 2, the phase difference between port 2 and 3 is equal to 0°. If the horizontal slot is moved upwards (see Fig. 4.2a) a phase difference of about −245° between port 2 and 3 is obtained (see Fig. 4.3). On the other hand, if the horizontal slot is moved downwards (see Fig. 4.2b) the phase difference is about −125° (see Fig. 4.3). The diodes are placed only at one radiator, to reduce losses.

As already mentioned at the beginning of this section, two radiators separated by $\frac{1}{2}\lambda_0$ (at 2.6 GHz) are used to generate reconfigurable radiation patterns. Due to their very compact dimension of 9.8 mm and 7.3 mm in length and height respectively, inverted-F antennas (IFA) were chosen for this design. Both

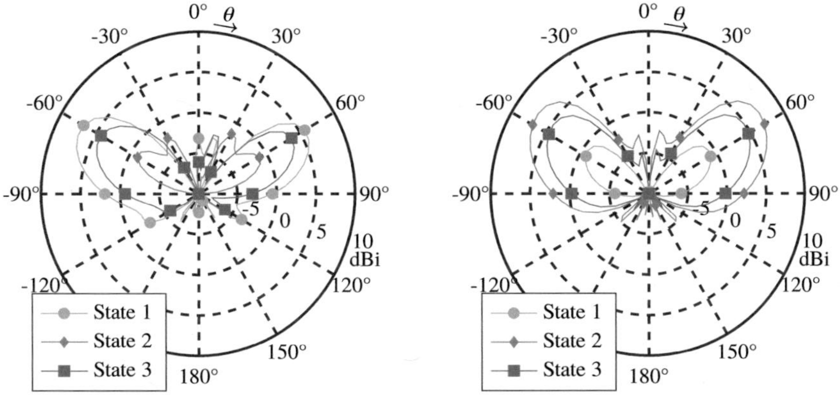


(a) Elevation cut for $\psi$=0°, $xy$-plane

(b) Elevation cut for $\psi$=90°, $yz$-plane

Figure 4.4: Simulated radiation pattern of the antenna using back-to-back balun for three reconfigurable states at 2.6 GHz.

antennas are placed at the top of 15 mm high vertical metal ground plane which is in turn mounted on a horizontal ($xz$ plane) ground plane. The overall antenna size is $66.2 \times 26.3\,\text{mm}^2$. Since the diodes are contrariwise polarized, they can be easily controlled by the polarity of the DC signal, which is fed through a line leading underneath the horizontal ground plane (see Fig. 4.1). Due to the construction of the feeding network, the antenna input port is galvanically isolated from DC signal and no additional DC block is needed.

In state 1 the two IFAs are fed with a signal in opposite phase (180° phase difference) resulting in two beams pointing in ± driving direction (see Fig. 4.4a). In this case the upper diode is ON and and the lower one is OFF. The maximum gain is 6.9 dBi in $-x$ direction and 5.7 dBi in $+x$ direction. The HPBW in azimuth is 80° and 137° in and against driving direction respectively, and the HPBW in elevation is 32°. The unequal feeding of the IFAs is caused by the proximity of the lower signal path to the ground plane. Some of the signal in the lower path is coupled to the ground plane and thus the impedance of the lower path changes. In order to activate state 2, the upper diode has to be OFF and the lower one ON, thus the radiators are fed with a signal in phase (0° phase difference). Two beams pointing towards the sides of the vehicle are generated (see Fig. 4.4b). In this case, the maximum gain is 7.2 dBi and the HPBW is 88° in azimuth, and

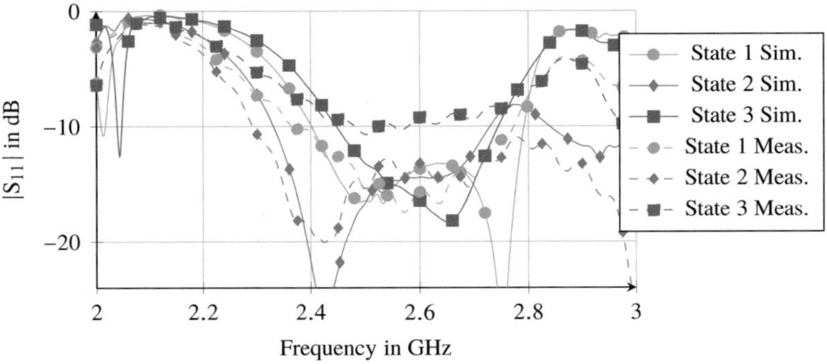


Figure 4.5: Simulated and measured S-parameters of the presented antenna using back-to-back balun for three reconfigurable states.

40° in elevation. For state 3 both diodes are OFF and thus only one IFA is fed. In this case a nearly omnidirectional pattern with maximum gain of 4.9 dBi is obtained (see Fig. 4.4).

Simulation results show that the antenna is matched between 2.35 GHz and 2.8 GHz for the three states, if −6 dB bandwidth is considered (see Fig. 4.5). The bandwidth is wider than expected for the IFA elements, however it has to be considered that in this case the feeding network improves the antenna matching, since the IFAs are not directly connected to a 50 Ω input port. Thus, the feeding network can also be seen as a matching network.

After finishing the design process, a prototype of this antenna was fabricated. The antenna structure was etched on a 0.787 mm thick Rogers RT5880 substrate. The p-i-n diodes (BAP64-02) and the resistor were soldered manually to it and the antenna was attached to a metal plated FR4 board ($150 \times 100\,\mathrm{mm}^2$) used as a ground plane. During the measurements in an anechoic chamber this structure was attached to a bigger ground plane ($500 \times 500\,\mathrm{mm}^2$).

The measured return loss for all reconfigurable states of the prototype is depicted in Fig. 4.5. Even though the matching is slightly worse than in the simulation, the bandwidth of the antenna is still bigger than 200 MHz (considering the −6 dB bandwidth). The biggest difference between simulation and measurement can be observed for state 3. Next, the radiation patterns and gain of the fabricated prototype were measured. The form of the measured pattern for the state 1

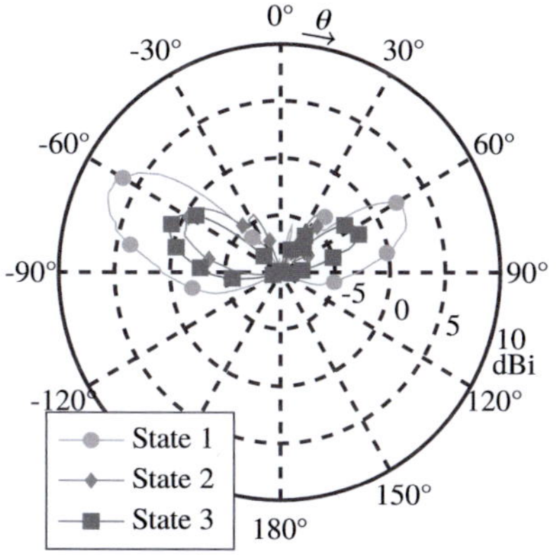


(a) Elevation cut for $\psi$=0°, $xy$-plane

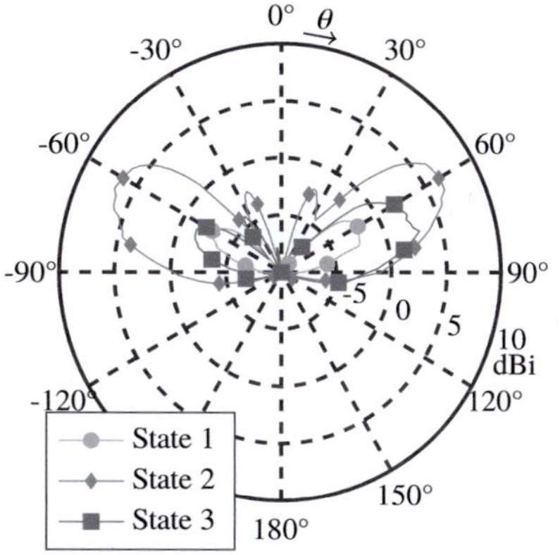


(b) Elevation cut for $\psi$=90°, $yz$-plane

Figure 4.6: Measured radiation pattern of the antenna using back-to-back balun for three reconfigurable states at 2.6 GHz.

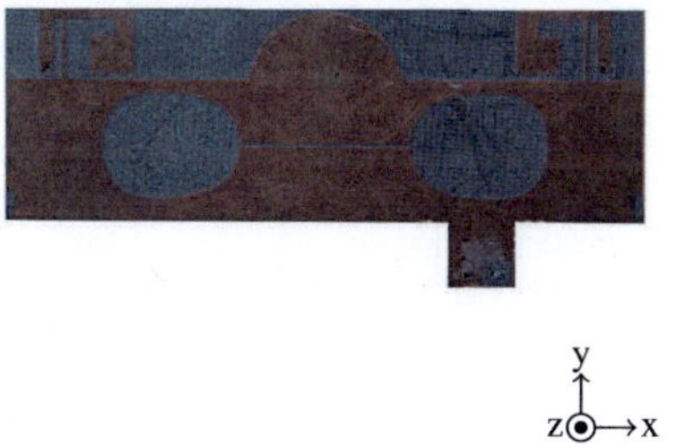


(a) Front side

(b) Back side

Figure 4.7: Fabricated prototype of the single antenna using back-to-back balun [14].

corresponds very well with the simulation (see Fig. 4.6a). The maximum gain in $\psi$=180° direction is about 6.9 dBi and thus the same as the simulated gain. In the $\psi$=0° direction the difference between the measured and simulated gain is 2 dB. The reason for the bigger difference in this case are inaccuracies in p-i-n diode modeling resulting in beam shift. For state 2 the maximum gain in $\psi$=90° is about 6.5 dBi (see Fig. 4.6b) and thus lower by about 0.7 dB than the simulated gain. As expected from the return loss measurement of state 3,

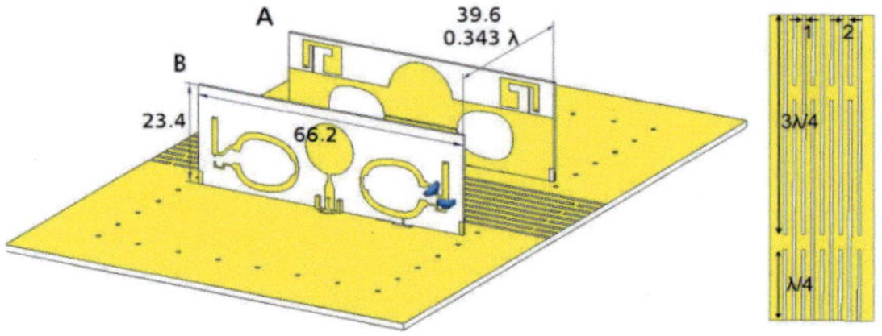


(a) (b)

Figure 4.8: Configuration of MIMO system (a) and decoupling slots (b). Dimensions are given in mm [14].

a significant difference between simulated and measured gain in both elevation cuts appears. For azimuth $\psi$=90°, a moderate gain of 3.5 dBi is observed (see Fig. 4.6b). The shape of the curve shows strong asymmetries between $\psi$=90° and 270°. The simulation did not show such asymmetries. In contrary such asymmetry can be observed in simulation for the $\psi$=0° cut. Also in this case the discrepancy between simulated and measured results occurs, and the maximum gain is 1.5 dBi.

### 4.1.2 MIMO Configuration

The presented reconfigurable antenna is then taken as a basis for the MIMO configuration. The antenna is mirrored along the $x-$axis and the second one is placed parallel to the first one in a distance of $0.343\lambda_0$ (see Fig. 4.8a). The overall size of this antenna system is $66.2 \times 39.6 \times 23.4\,\mathrm{mm}^3$ and its fractional bandwidth is 17 %. The two antennas face each other with the ground plane sides where the slotline is etched. The reason for this configuration is that simulations show a lower amplitude of the E-field on this side of the antenna and thus, the mutual coupling between antennas is decreased. The two antennas are referred to as A and B (see Fig. 4.8a). A set of decoupling slots of $\frac{1}{4}\lambda_0$ and $\frac{3}{4}\lambda_0$ is placed between the antennas (see Fig. 4.8b), which decreases the coupling between the antennas by modest 1.5 dB. It is due to the fact that most of the energy is coupled by radiated fields. Nevertheless, the envelope correlation coefficient (ECC) for the combination of all states of both antennas has a maximum value of ECC of about 0.07, making the system appropriate for MIMO applications.

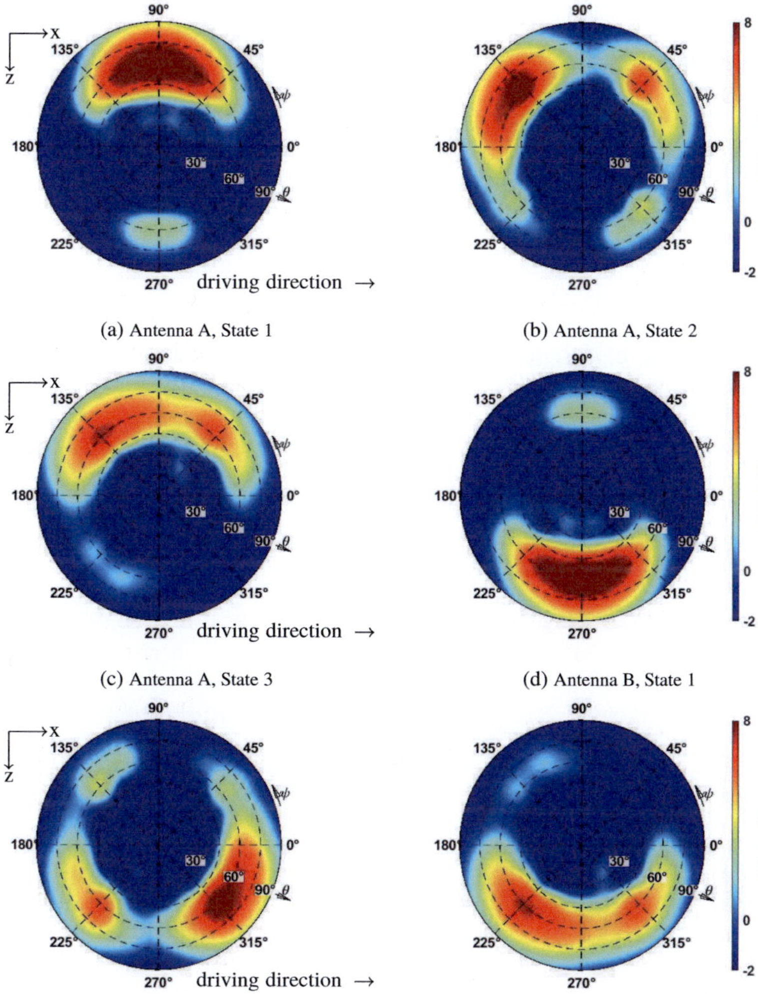


(a) Antenna A, State 1 (b) Antenna A, State 2

(c) Antenna A, State 3 (d) Antenna B, State 1

(e) Antenna B, State 2 (f) Antenna B, State 3

Figure 4.9: Simulated radiation patterns of antenna A and B for the presented MIMO configuration at 2.6 GHz.

Table 4.1: Radiation pattern parameters of the presented MIMO antenna system using back-to-back balun.

| State | Gain in dBi | Beam direction $\psi$ | HPBW$_{\psi}$ |
|---|---|---|---|
| A1 | 8.7 | 90° | 90° |
| A2 | 8.5 | 45° and 135° | 60° |
| A3 | 7.6 | 90° | 138° |
| B1 | 8.7 | 270° | 90° |
| B2 | 8.5 | 225° and 315° | 60° |
| B3 | 7.6 | 270° | 138° |

The ECC value is calculated from antennas' s-parameters based on the formula (4.1) given in [BRC03]:

$$\rho_e = \frac{|S_{11}^* S_{21} + S_{21}^* S_{22}|^2}{(1 - (|S_{11}|^2 + |S_{21}|^2))(1 - (|S_{22}|^2 + |S_{12}|^2))}. \qquad (4.1)$$

Where $S_{11}$ and $S_{22}$ are the return losses of antennas A and B respectively, and $S_{21}$ and $S_{12}$ is the isolation between the antenna ports.

Looking at the radiation patterns, it can be observed that in used parallel configuration the antennas shadow each other (see Fig. 4.9). It can especially be observed for state 1 (see Fig. 4.9a and Fig. 4.9d). Nevertheless, the angular regions covering the ± driving direction and the other orthogonal to it remain unchanged. Compared to the gain of a single antenna, an increase by 1.7 dB, 0.6 dB and 2.5 dB can be observed for states 1, 2 and 3 respectively. The summary of the most important radiation parameters of the presented antenna system is given in table 4.1.

## 4.2 A Low-Profile Multiple ESPAR Antenna System

Another approach discussed in the previous chapter is the parasitic array antenna. In this section a compact antenna system with pattern reconfigurability at 2.6 GHz is presented [2]. The antenna height is reduced by top loading. As

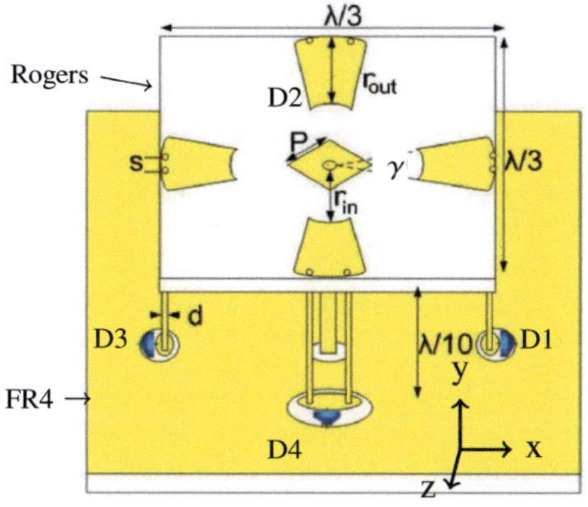


Figure 4.10: Perspective view of the ESPAR antenna [2].

a result of the small distance between the antennas ($\frac{1}{3}\lambda_0$ at 2.6 GHz) a very compact antenna with an overall size of $85 \times 45 \times 13.2\,\mathrm{mm}^3$ is obtained. Each of the antennas can generate six reconfigurable patterns. The design of a single antenna, the choice of a decoupling structure and the MIMO antenna system are presented in this section.

### 4.2.1 Single Element

The single ESPAR antenna consists of a primary radiator and parasitic elements placed around it (see Fig. 4.10). In case of the presented design, top loaded monopoles were chosen to decrease the antenna height. Thus, the height could be reduced from $\frac{1}{4}\lambda_0$ to $\frac{1}{10}\lambda_0$ (at 2.6 GHz).

The primary radiator consists of a driven post of height $h$ which is 11.5 mm and a rectangular metallic hat (see Fig. 4.10) with a diameter $p$ equal to 7.2 mm as capacitive top loading (a detailed study of top loading choice is presented in [2]). The radiator is fed directly by a coaxial line. Due to the top loading, the current on the vertical post has a nearly uniform distribution, and this part of the antenna is responsible for the radiation. The current on the top-hat decreases rapidly from the center to the edge. By increasing the size of the top-hat the antenna resonant frequency moves towards lower frequencies. Thus, increasing the top-hat radius effectively reduces the antenna height. At the same time the real part of the impedance decreases when the antenna height is reduced. Therefore, a compromise is needed between antenna height and a good matching. The use of an additional shorting post establishing a folded monopole and thus increasing the impedance is not necessary since the parasitic elements increase the input impedance.

Directive reconfigurable patterns can be obtained by placing parasitic elements in the vicinity of the driven element (see Fig. 4.10). The parasitic elements are also top loaded monopoles and consist of vertical posts with a top-hat. The top-hat differs in its form from the one used for the primary radiator. A sector-ring shape is chosen for the parasitic elements as it offers a more compact antenna size than the rectangular shape. Tuning the element to the required resonant frequency involves changing the angle $\gamma$ and the radius $r_{out}$. Therefore, the radius can be kept small, while increasing the angle $\gamma$, resulting in reduced antenna dimensions. Based on simulation results, the parameters are set to $\gamma = 17°$ and $r_{out} = 10\,\text{mm}$ to achieve reflector behavior of the parasitic elements (lower resonant frequency than that of the driven element). Another important design issue is the number of the vertical posts. The parasitic elements use two vertical posts per element. Introducing a second post adds inductance and capacitance, thus reducing the size of the element further. Moreover, unlike in case of the primary radiator, the posts are placed on the outer side and not in the middle of the element. Thus, the resonant frequency is reduced and additionally antenna's efficiency increases.

The distance between the parasitic element and the driven element contributes to size reduction as well. Reducing the distance $r_{in}$ increases the coupling between the driven and parasitic element causing lowering of the resonant frequency. However, this causes also a reduction in the antenna performance in terms of efficiency and gain. As a trade-off between antenna size and performance $in_r$ is set to 12 mm. Thus, the distance between the phase center of the primary radiator and the one of a parasitic element is about $0.14\lambda_0$.

The patterns optimized for automotive antennas can be realized by using four parasitic elements. A larger number of parasitic elements does not increase the antenna's flexibility, but decreases the antenna efficiency. A parasitic element acts like a reflector when shorted to ground and acts like a director when left "floating". Therefore, the elements have to be connected to ground with electrical switches, in this case p-i-n diodes. For each element an elliptical slot

Table 4.2: Switch combinations and beam directions for the different reconfigurable states of the ESPAR.

| State | D1 | D2 | D3 | D4 | Beam direction $\psi$ |
|---|---|---|---|---|---|
| 1 | ON | ON | OFF | ON | 180° |
| 2 | OFF | ON | ON | ON | 0° |
| 3 | ON | OFF | ON | ON | 90° |
| 4 | ON | ON | ON | OFF | 270° |
| 5 | ON | OFF | ON | OFF | 0° and 180° |
| 6 | OFF | ON | OFF | ON | 90° and 270° |
| 7 | ON | ON | OFF | OFF | 225° |
| 8 | OFF | ON | ON | OFF | 315° |
| 9 | OFF | OFF | ON | ON | 45° |
| 10 | ON | OFF | OFF | ON | 135° |

in the ground plane is etched and a metallic pad is placed in the middle of each slot. The wires are connected to this pads as shown in Fig. 4.10. A p-i-n diode placed between the pad and the ground is used at each element. When the diode is switched ON, the current on the corresponding shorting wires is around −80° out of phase with the current flowing in the radiator (reflector function). If the diode is switched OFF, the parasitic element is around 140° out of phase with the radiator (director function).

The DC feeding network is placed underneath the ground plane. It consist of metallic pads connected to the vertical posts, metallic strips and 5 kΩ resistors as RF choke. The ground plane and DC network are printed on FR-4 substrate with permittivity $\varepsilon_r = 4.3$ and a thickness of 1.6 mm. Whereas the top-hat and parasitic elements are printed on Rogers RT5880 substrate with a thickness of 0.508 mm.

The antenna can generate 10 different patterns. The possible switch configurations are presented in the table 4.2. As discussed in the introduction the antenna patterns should cover the driving direction ($+x$), the opposite direction ($-x$) and directions to the left and right of the car ($\pm y$ direction). These patterns are realized by states 1–4 and achieve maximum simulated gain of 6.9 dBi. The front-to-back ratio (FBR) is 3.5 dB. In some cases it might be important to cover the ± driving direction or left/right direction simultaneously. These scenarios

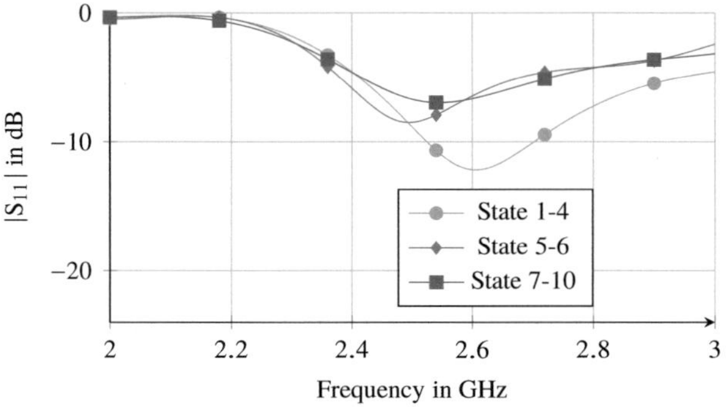


Figure 4.11: Simulated S-parameters of the presented ESPAR for different reconfigurable states.

are covered by states 5 and 6 with a maximum gain of 5.7 dBi. Additionally, the antenna can generate patterns directed towards the corners of the car roof $\psi = 45°, 135°, 225°$ and $315°$ (left-front, left-rear, right-rear, right-front). These directions are very important in urban scenarios as discussed in section 2.8. These patterns are realized by states 7 to 10 with a maximum gain of 7.2 dBi. The FBR for these states is 6.1 dB. The simulated total antenna efficiency has values between 70 % for states 7 to 10 and 77 % for states 5 and 6.

The best matching is obtained for states 1 to 4 (2.43 GHz to 2.86 GHz), in which case three diodes are switched ON (see Fig. 4.11). The curve is also shifted towards higher frequencies compared to the curves for states 5 and 6 (2.4 GHz to 2.62 GHz), and 7 to 10 (2.45 GHz to 2.66 GHz). In general, a worse matching is observed for states in which two diodes are in ON state. Looking at the simulation we can see stronger currents flowing between the floating elements (OFF state) and the antenna port, therefore diminishing the matching for states 5 to 10.

## 4.2.2 MIMO Configuration

A MIMO antenna system consists of two reconfigurable antennas as presented in section 4.2.1. The antennas share a common ground plane and the distance between the phase centers of both antennas is about $\lambda_0/3$ along the driving direction (see Fig. 4.12). The small distance was chosen in order to obtain a

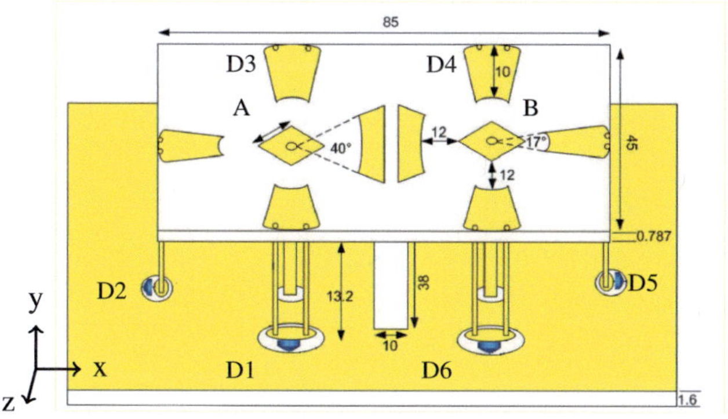


Figure 4.12: Perspective view of the low-profile MIMO antenna [2]. Dimensions are given in mm.

compact design. The trade off in this case is higher coupling between the antennas. The reconfiguration of adjacent parasitic elements positioned between the driven elements of both antennas becomes useless, since the antennas shadow one another. Thus, these elements are designed to act as reflectors, and the vertical posts are removed to reduce the coupling between these elements. Therefore, the size of these two parasitic elements is increased so that they are still able to act as a reflector. Furthermore, by decreasing their radius it becomes possible to place the antennas closer together. A growth in size of this elements is achieved by increasing the angle of the arc section of the element which is now $\gamma = 40°$, more than twice the original value (see Fig. 4.12).

Due to the small distance between the antennas, a strong coupling can be observed. This leads to a decrease in the antennas' radiation efficiency. Therefore, a decoupling structure should be used to solve this problem. The dominant role in mutual coupling is played by currents on the ground plane. Various techniques reducing mutual coupling by modifying ground plane structure are known from literature [CCMR07, ZXX09, COII04]. Three different concepts (see Fig. 4.13) were tested and compared.

The H-shaped decoupling structure consists of slots of width $w$ interleaved with metallic strips of width $s$ and length $k$. The elements are placed in the middle of the ground plane (see Fig. 4.13a). The structure can be modeled as capacitance (for metallic stubs) and inductance (connecting strip of width $c$). Thus, it is equivalent to a parallel LC network. If the values are set correctly the

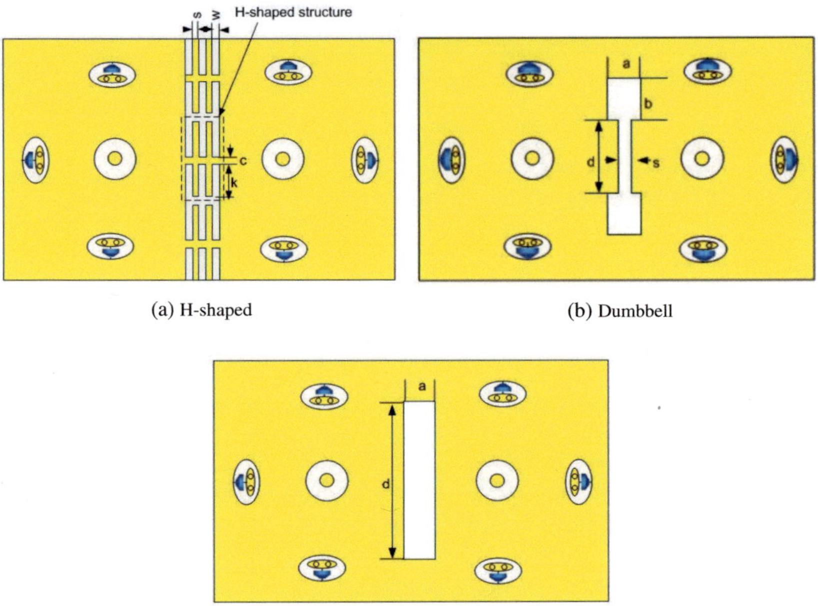


(a) H-shaped

(b) Dumbbell

(c) Slotted ground plane

Figure 4.13: Evaluated decoupling structures [2].

LC network works as a band-stop filter blocking the currents between elements at the desired frequency. If the design parameter values are set to $k$ = 19 mm, $s$ = 0.5 mm, $w$ = 0.5 mm, $c$ = 1.5 mm, the band-stop region from 2.2 to 2.75 GHz with a mutual coupling across the band lower than −10 dB is achieved (the structure was tested by placing it on FR4 substrate and connecting it to a microstrip line with two ports in an EM simulator).

The second option is a dumbbell slot, etched in the ground plane as shown in Fig. 4.13b. The dumbbell rectangular head has a length $a$ and width $b$, and the length and width of the slot are $d$ and $s$ respectively. The dumbbell structure reduces mutual coupling by suppressing the surface wave if the resonant frequency of the MIMO antenna falls in its band gap. Increasing $a$, $b$ or $d$ reduces the band-stop frequency, while $s$ should be decreased to lower the frequency. The design parameter values are set to $a$ = 5 mm, $b$ = 5 mm, $d$ = 18 mm, $s$

= 1 mm. The dumbbell was tested using the same procedure as for H-shaped structure. For the determined parameters band-stop region from 2 to 2.9 GHz with a mutual coupling across the band lower than −10 dB is achieved.

The last option is a slotted ground plane (SGP) (see Fig. 4.13c). It is designed to resonate at nearly the design frequency of the antenna. The currents flowing between the antennas are captured in the slot and thus the mutual coupling is reduced. The slot width $a$ is 10 mm and the length $d$ is 38 mm. Thanks to its width, the structure achieves a band-stop characteristic within a wide band from 1.43 to 3.45 GHz (at least isolation −10 dB).

The decoupling structures improve the isolation between the ports but at the same time they reduce the performance of the antenna in terms of radiation efficiency. Reduced radiation efficiency is the effect of capturing the fields in the decoupling structures. Since these fields would normally contribute to antennas' radiation, the efficiency drops. Thus, a compromise between coupling reduction and antenna gain has to be found. The resonance frequency of the decoupling structure is placed below the center frequency of the MIMO system for the best performance. Out of all evaluated structures, SGP offers the best isolation of around 25 dB and lowest influence on antenna radiation.

The mechanism of the p-i-n diode switching for a MIMO antenna is the same as that used for a single antenna. Thus, switching each p-i-n diode ON connects the parasitic element to the metallic ground plane. Since the shorting post of the inner parasitic elements were removed, one diode per antenna is eliminated, resulting in three diodes per antenna. The two antennas are 180° symmetric in azimuth and so are the possible reconfigurable states. The configuration of the diodes is shown in Fig. 4.12. Diodes D1, D2, D3 are used for pattern reconfiguration of antenna A whereas diodes D4, D5, D6 are used for antenna B. Resulting from this configuration six reconfigurable states per antenna are possible (see Fig. 4.14). The MIMO antenna patterns cover the relevant directions like ± driving direction, right-front, left-front, right-rear, left-rear, or left/right direction. The highest gain is obtained by antennas A and B in state 1 (± driving direction) and it is about 9.2 dBi. The lowest gain is observed for state 6 of both antennas (a gain of 4.6 dBi in left/right direction). Due to the small distance between the antennas, the active parasitic element of the adjacent antenna affects the radiation pattern of the considered antenna. A shift in the main beam radiation direction in azimuth from 90° to around 120° can be observed for

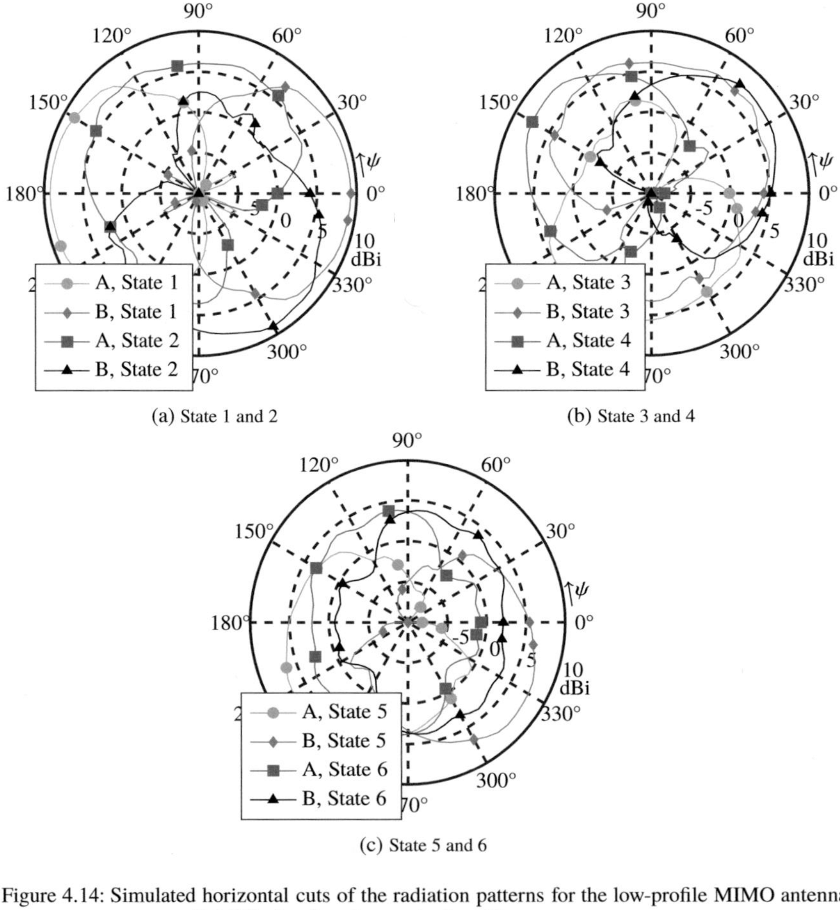


Figure 4.14: Simulated horizontal cuts of the radiation patterns for the low-profile MIMO antenna system at 2.6 GHz for elevation angle $\theta$ =60°.

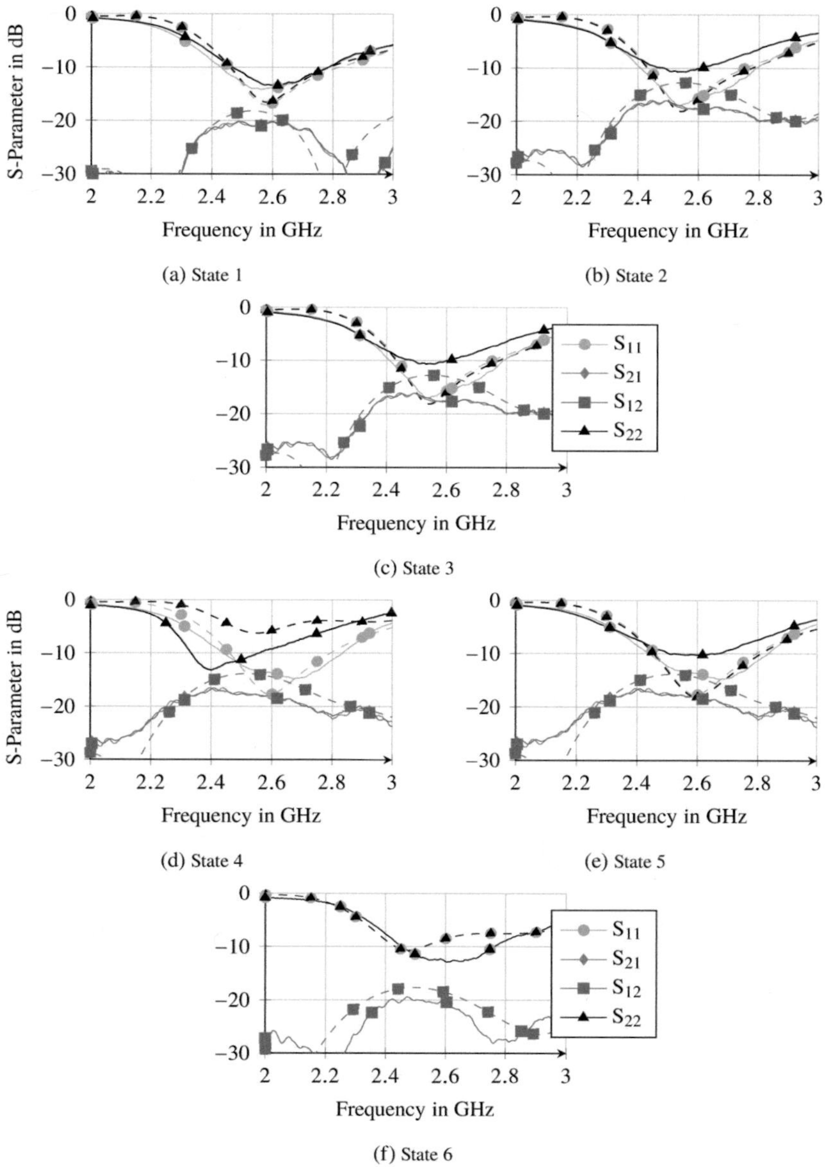


(a) State 1

(b) State 2

(c) State 3

(d) State 4

(e) State 5

(f) State 6

Figure 4.15: Simulated (dashed) and measured (solid) S-parameter for the MIMO antenna.

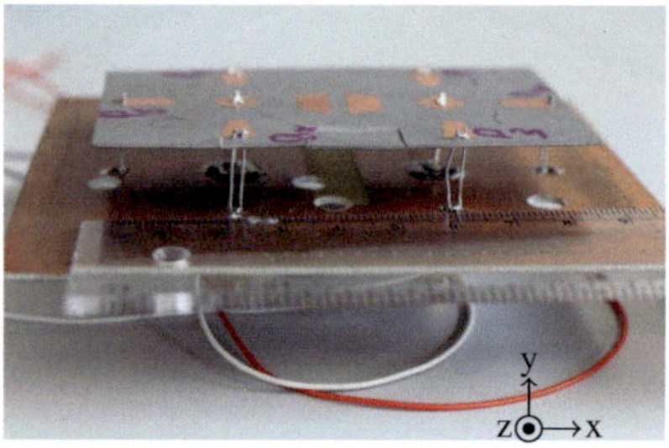


(a) Perspective view

(b) Bottom side

Figure 4.16: Fabricated prototype of the low-profile MIMO antenna [2].

antenna A in state 2 (see Fig. 4.14a). This is due to the active parasitic element D4 of the second antenna. A similar shift is also observed for states 3 to 6.

The S-parameters of the presented antenna are shown in Fig. 4.15. The matching of about −17 dB is achieved with the decoupling structure (SGP) as compared to −10 dB without decoupling structure. The bandwidth at −6 dB return loss covers the frequency range from 2.4 to 2.8 GHz for all states (15 % fractional bandwidth). The matching for states 4, 5 and 6 is worse than for the other states. This is because two diodes per element are OFF and as already discussed stronger currents flow from the floating parasitic elements to the antenna port. All the states show a good isolation of at least −14 dB at the design frequency (see Fig. 4.15).

In the next step a prototype of the antenna (see Fig. 4.16) was fabricated and measured. As before the p-i-n diode used is BAP64-02 from NXP. The measured input reflection coefficient of antennas A and B in different reconfigurable states corresponds well with the simulation results (see Fig. 4.15). Considering a matching better than −6 dB for the antenna bandwidth, the antenna covers the frequency range from 2.4 to 2.8 GHz. The measurement results show a very good isolation in all states (see Fig. 4.11). Its value is slightly better than the simulated one (see Fig. 4.16) which can be partially explained by increased losses in the prototype compared to the EM simulation model.

In the following step the calibrated gain patterns were measured inside an anechoic chamber. For this purposes the prototype was attached to a large ground

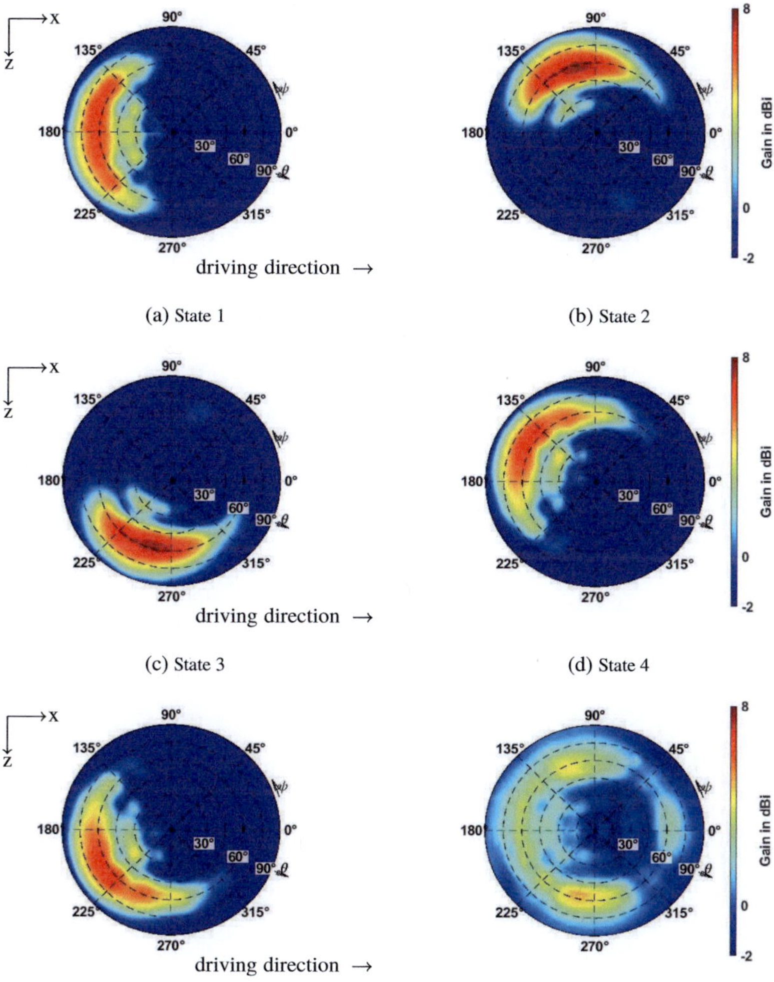


(a) State 1 (b) State 2

(c) State 3 (d) State 4

(e) State 5 (f) State 6

Figure 4.17: Measured radiation patterns of antenna A for the presented low-profile MIMO antenna system at 2.6 GHz.

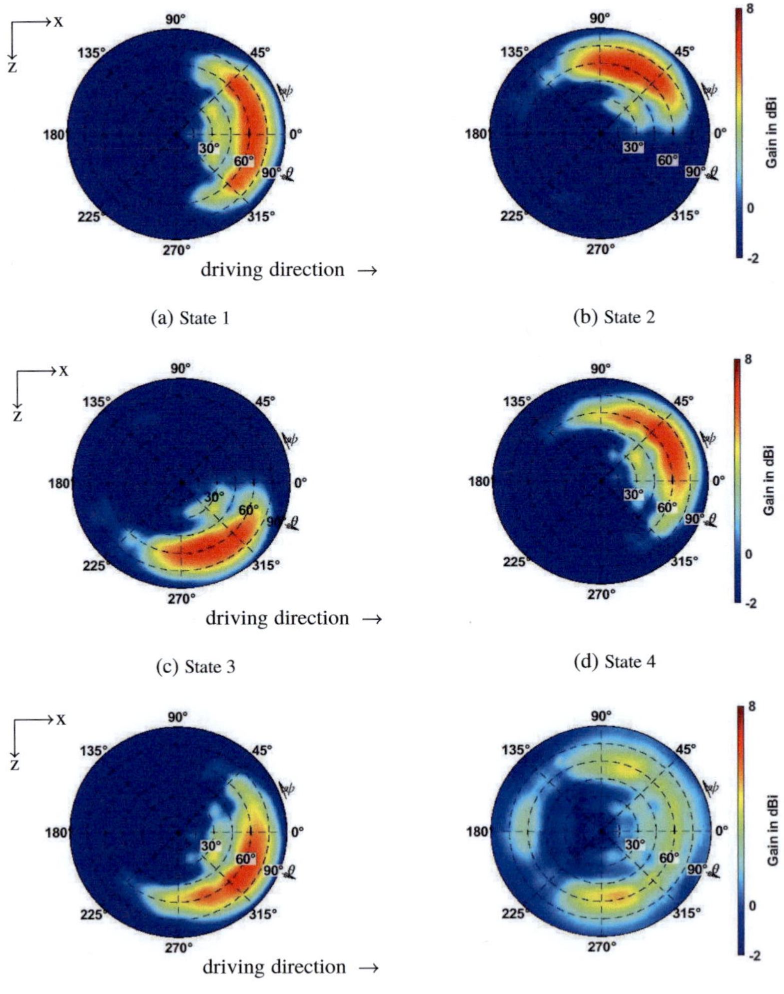


(a) State 1

(b) State 2

(c) State 3

(d) State 4

(e) State 5

(f) State 6

Figure 4.18: Measured radiation patterns of antenna B for the presented low-profile MIMO antenna system at 2.6 GHz.

Table 4.3: Radiation pattern parameters of the presented low-profile MIMO antenna system.

| State | Gain in dBi | Beam direction $\psi$ | $HPBW_{\psi}$ |
|---|---|---|---|
| A1 | 7.6 | 180° | 105° |
| A2 | 7.9 | 105° | 80° |
| A3 | 8 | 255° | 80° |
| A4 | 7.2 | 140° | 110° |
| A5 | 7.1 | 243° | 110° |
| A6 | 4 | 100° and 260° | 240° |
| B1 | 7.4 | 0° | 105° |
| B2 | 7.8 | 285° | 80° |
| B3 | 7.9 | 75° | 80° |
| B4 | 7.1 | 40° | 110° |
| B5 | 7.2 | 320° | 110° |
| B6 | 4.1 | 80° and 280° | 240° |

plane with a size of $500 \times 500\,\text{mm}^2$. The drill holes visible in the Fig. 4.16 are necessary in order to mount the antenna on the rotary tower inside the anechoic chamber. The measured patterns are presented in Fig. 4.17. Comparing the simulated and measured gain values, the biggest difference can be observed for state 1 where instead of the simulated 9.2 dBi gain 7.4 dBi is measured (see Fig. 4.18a and 4.14a). Nevertheless, it can be observed that the measurement results correspond very well with the simulation results. Not only similar gain values are obtained for the remaining states but also the form of the patterns corresponds well. The summary of the most important radiation parameters of presented low-profile antenna system is given in table 4.3.

In the next step the ECC values for representative antenna states were calculated based on equation 4.1. All values are below 0.015 making the antenna attractive for MIMO applications.

## 4.3 A Compact Multiple Antenna System

Yet another approach to realize the desired patterns is to use two pattern reconfigurable antennas, as presented in section 3.5.2, to constitute a multiple antenna system [22]. The single antenna consists of two radiators separated by $\frac{3}{8}\lambda_0$ (at 2.5 GHz), which are fed in parallel. However, unlike in the case of the multiple antenna system from section 4.1, where the single antennas are placed parallel to each other, the antennas here are perpendicular to each other (see Fig. 4.19). Thus, the effect of antennas shadowing each other can be minimized. Furthermore, there are two additional parasitic elements ($P_1$ and $P_2$) introduced at one of the single antennas, adding two reconfigurable patterns. The overall antenna size is $63 \times 65 \times 18\,\text{mm}^3$ and its fractional bandwidth is 8 %.

### 4.3.1 Parasitic Elements

In order to improve the directivity of the switchable patterns generated by the antenna and to introduce additional reconfigurable patterns with higher gain, parasitic elements are used. The choice of elements is based on the ability to switch between a reflector and a director mode. A possible solution is an U-slot reflector which is used commonly in chipless RFID (radio frequency identification) technology to create frequency signatures in the back scattered signal [IK15]. In order to act as a reflector, the slot should have a total length of about $\lambda_0/2$ at the frequency of interest, which corresponds to 60 mm at the design frequency. Since the parasitic element is printed on a FR-4 substrate with an $\varepsilon_r$ of 4.3 and the p-i-n diodes used in the design provide some additional inductance, the physical length is reduced to approximately 35 mm (16 mm long horizontal slot and two 9.5 mm long vertical slots with 1 mm width). The U-slot is placed on a metallic rectangular patch with 24 mm by 18 mm width and height respectively (see Fig. 4.19b). The p-i-n diodes are mounted at the ends of the horizontal slot, where additional slots of 0.5 mm width introduce a separation in the metallization. The whole structure is then mounted perpendicularly to the ground plane. To generate the back scattered signal the diodes have to be in ON state and the E-field is excited in the whole U-slot (the diodes connect the separated metallic parts of the patch). In this case the simulated ratio of the back scattered field to the forward scattered field is about 10 dB. In order to switch the element to director mode, the diodes have to be switched OFF. Thus, the E-field is excited only in a horizontal slot and on the edges of the patch. In this case the simulated ratio of back scattered field to forward scattered field is about −4.5 dB and the wave is directed.

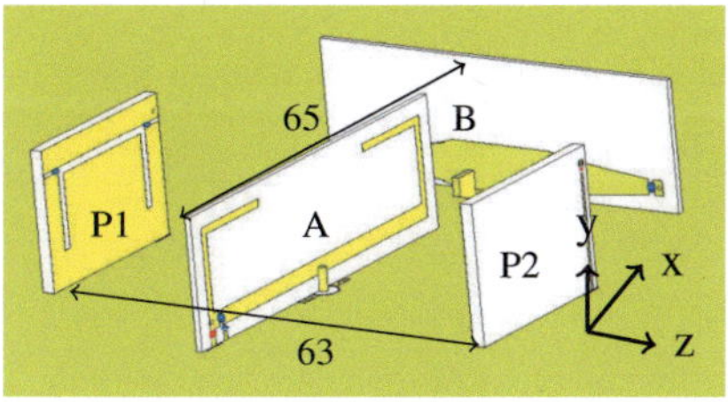


(a) Perspective view

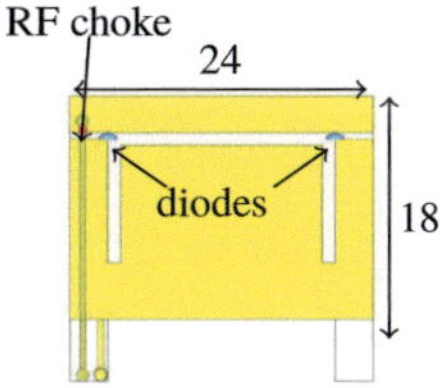


(b) Parasitic element

Figure 4.19: Perspective view of the compact multiple antenna system and layout of the parasitic element. Dimensions are given in mm [22].

### 4.3.2 MIMO Configuration

As already mentioned, the unit antennas of the MIMO system are placed orthogonal to one another (see Fig. 4.19a). Parallel arrangement of the antennas was considered as well. However, in this case a shadowing effect appears blocking the beams in direction of the second antenna. Thus, the required patterns cannot be obtained. Therefore, the orthogonal arrangement is chosen. Size is a very important design parameter for automotive antennas, since they should fit within the commonly used shark fin housing. The horizontal distance between the phase centers of the antennas is 40.5 mm ($\frac{1}{3}\lambda_0$) and is chosen to preserve small footprint and compact dimension of the system. However, the coupling between the antennas for the chosen separation is still acceptable. The antennas are defined as antenna A and antenna B (see Fig. 4.19a). The parasitic elements are placed along the axis orthogonal to the substrate of antenna A ($x$-axis) in the distance of 30 mm from antenna A, which is $\frac{1}{4}\lambda_0$ at design frequency. This distance proves to be optimal in terms of radiation pattern and antenna matching.

The mechanism of pattern switching for a single antenna was explained in detail in section 3.5.2. For antenna B there is no difference compared to the antenna from section 3.5.2 and for antenna A the two first states remain the same. Therefore, we will concentrate on the two additional states of antenna A. There are three switches (p-i-n diodes) on each antenna: D1, D2, D3. Each of the parasitic elements is provided with two diodes $D4_1$ and $D4_2$ for element P1, and $D5_1$ and $D5_2$ for element P2.

Table 4.4: Configuration of switching elements and control voltage

| State | D1 | D2 | D3 | D4 | D5 | $V_D$ | $V_{bias-T}$ | $V_{P1}/V_{P2}$ |
|---|---|---|---|---|---|---|---|---|
| 1 | ON | OFF | OFF | OFF | OFF | −10 V | 0 V | 0 V/0 V |
| 2 | OFF | ON | ON | ON | ON | 10 V | 1 V | −10 V/−10 V |
| 3 | ON | OFF | OFF | OFF | ON | −10 V | 0 V | −10 V/0 V |
| 4 | ON | OFF | OFF | ON | OFF | −10 V | 0 V | 0 V/−10 V |

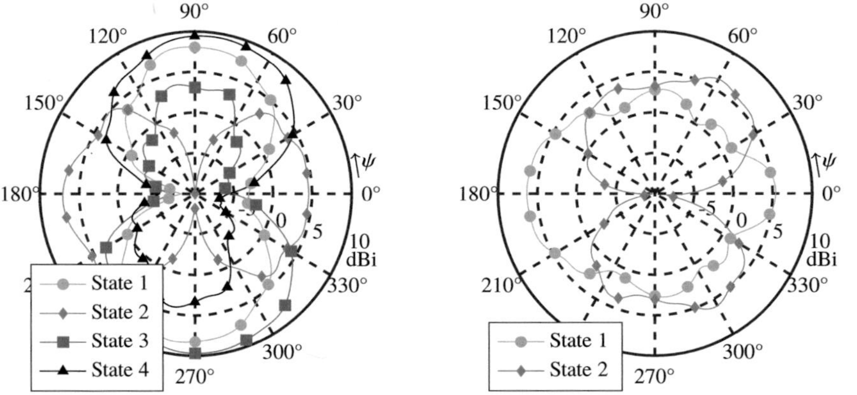


(a) Antenna A (b) Antenna B

Figure 4.20: Simulated horizontal cuts of the radiation patterns for the compact MIMO antenna at 2.5 GHz for elevation angle $\theta$ =60°.

For the proper realization of states 1 and 2 of antenna A, it is crucial that the diodes on the parasitic elements are OFF for state 1 and ON for state 2. For these states antenna A obtains a maximum gain between 7.3 dBi and 7.7 dBi and antenna B between 6.1 dBi and 7.8 dBi at 2.5 GHz (see Fig. 4.20). To activate state 3, the diodes on the primary radiator are configured the same as for state 1. Thus, beams in $\pm z$ direction are created (to the vehicles left and right). If the diodes on the parasitic element P1 are ON and those on the other parasitic element are OFF, the beam is directed in $+z$ direction. Activating the diodes on element P2 and deactivating those on element P1 directs the beam in $-z$ direction. For these states a maximum gain value is between 9.1 dBi and 9.7 dBi (see Fig. 4.20a).

Figure 4.21: Fabricated prototype of the compact MIMO antenna system [22].

Table 4.5: Comparison of maximal gain at 2.5 GHz for the different states of compact MIMO antenna system

| Antenna 1 | 1 | 2 | 3 | 4 |
|---|---|---|---|---|
| Simulation | 7.7 dBi | 7.3 dBi | 9.7 dBi | 9.1 dBi |
| Measurement | 7.7 dBi | 6.8 dBi | 8.4 dBi | 7.4 dBi |
| Antenna 2 | 1 | 2 | | |
| Simulation | 7.8 dBi | 6.1 dBi | | |
| Measurement | 8 dBi | 5.8 dBi | | |

After evaluating the performance of the antenna system by simulation a prototype of the antenna system was fabricated. Also in this case the BAP64-02 p-i-n diodes from NXP were used. The radiators were etched on 0.787 mm thick Rogers RT5880 substrate and the parasitic elements on 1.6 mm thick FR-4 substrate. The measured input matching is similar to results presented in the section 3.5.2. The matching for states 3 and 4 of antenna A is similar as for the state 1.

The measured gain for state 1 of antenna A and B is 7.7 dBi and 8 dBi respectively, and the values are similar to simulated ones (see table 4.5). For state 2 a maximum gain of 6.8 dBi and 5.8 dBi is measured (antennas A and B respectively). Also in this case differences between simulation and measurement are lower than 0.5 dB. The difference between simulation and measurement is higher for states 3 and 4 of antenna A and gain values of 7.4 dBi and 8.4 dBi are measured. The difference between these two states might be caused by fabrication inaccuracies, which result in different scattering from the parasitic elements P1 and P2. Nevertheless, the form of the different patterns corresponds very well

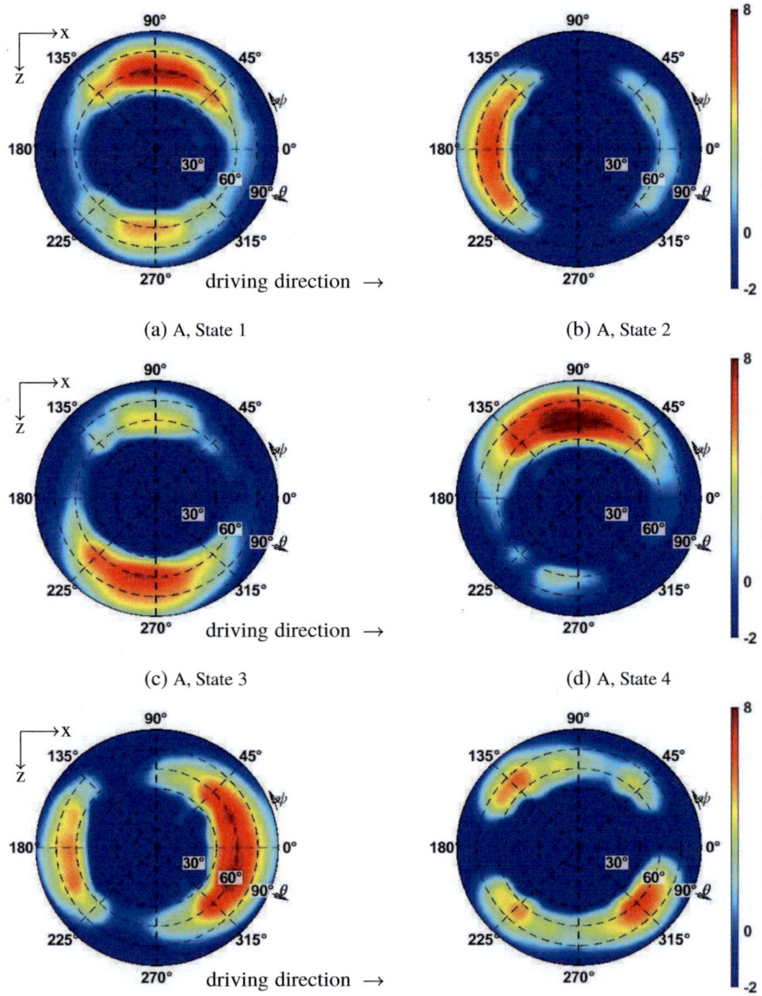


(a) A, State 1 (b) A, State 2

(c) A, State 3 (d) A, State 4

(e) B, State 1 (f) B, State 2

Figure 4.22: Measured radiation patterns of the presented compact MIMO antenna system at 2.5 GHz.

Table 4.6: Radiation pattern parameters of the presented compact MIMO antenna system.

| State | Gain in dBi | Beam direction $\psi$ | HPBW$_\psi$ |
|---|---|---|---|
| A1 | 7.7 | 90° and 265° | 100° |
| A2 | 6.8 | 0° and 180° | 90° |
| A3 | 8.4 | 260° | 100° |
| A4 | 7.4 | 100° | 100° |
| B1 | 8 | 0° and 180° | 120° |
| B2 | 5.8 | 135° and 315° | 110° |

with simulation results (see Fig. 4.22). The most important radiation parameters of this MIMO antenna system are given in table 4.6.

The ECC values for this antenna system, calculated based on equation 4.1 are below 0.08 making the antenna attractive for MIMO applications. Furthermore, the antenna system realizes patterns with high gain, making it a potential candidate for channel capacity enhancement.

## 4.4 A Wideband Multiple Antenna System

The designs presented so far in this chapter produced radiation patterns covering simultaneously the ± driving direction and and both sides of the vehicle simultanously. This requirement is mainly based on the results presented in [5]. However, as it has been presented in section 2.8 these requirements can be modified. Following the new requirements a compact wideband antenna system is proposed [21]. The single antenna is based on an element switching concept similar to the antenna discussed in section 3.3.2. With its overall size of $50.8 \times 30 \times 24\,\text{mm}^3$, this system is very compact and smaller by 28 % than the antenna system from section 4.2 and 50 % smaller than the antenna system from section 4.3. The small dimensions make it very attractive for automotive applications. Furthermore, thanks to a wideband matching from 2.1 to 2.87 GHz (31 % fractional bandwidth) this antenna system is very immune to frequency shifting due to fabrication imperfections and can potentially cover different services like various 4G frequency bands and Wi-Fi simultaneously.

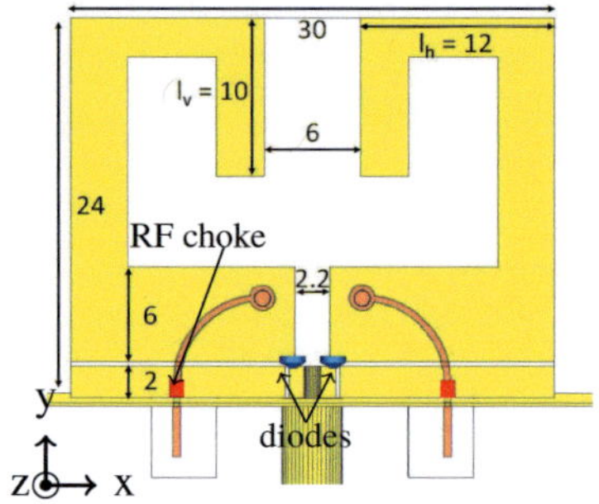


Figure 4.23: View of the single reconfigurable antenna. Dimensions are given in mm, yellow color represents the front side and orange the back side [21].

### 4.4.1 Single Element

The single antenna consists of two "C"–shaped radiators mirrored at the *y*-axis (see Fig. 4.23). The elements are tightly coupled and separated by only 2.2 mm in the lower part and by 6 mm in the upper part. The slot between the radiating elements and the ground plane of CPW feed is 0.3 mm wide. The radiators are connected to the central conductor of the CPW using p-i-n diodes. Thus, by switching the diodes three reconfigurable patterns can be generated. The DC feed used for biasing the diodes is placed on the backside of the antenna. The arc form of the DC lines and resistors embedded in these lines reduce the influence on the antenna's performance. Furthermore, the resistors serve as RF chokes. The vias connecting the DC feed to the radiators are placed in a section of the radiators where the lowest surface current at high frequencies was observed.

The choice of the "C"–shape for the radiators enables a wideband performance and compact design of the antenna. The antenna parts responsible for miniaturization are the upper horizontal bar and the vertical bar connected to it. By increasing the lengths $l_h$ and $l_v$ of horizontal and vertical bar respectively, the operation frequency can be decreased (see Fig. 4.24). However, increasing the length $l_v$ reduces the bandwidth. Furthermore, as the lengths are increased the antenna's gain and front-to-back ratio (FBR) are reduced. If $l_v$ is fixed to 10 mm and the value of $l_h$ is varied to 10 mm, 12 mm and 14 mm, the simulated gain value at 2.5 GHz is 6.6 dBi, 5.8 dBi and 5.4 dBi respectively. Likewise if $l_h$ is fixed to 12 mm and the value of $l_v$ is varied to 8 mm, 10 mm and 12 mm, the

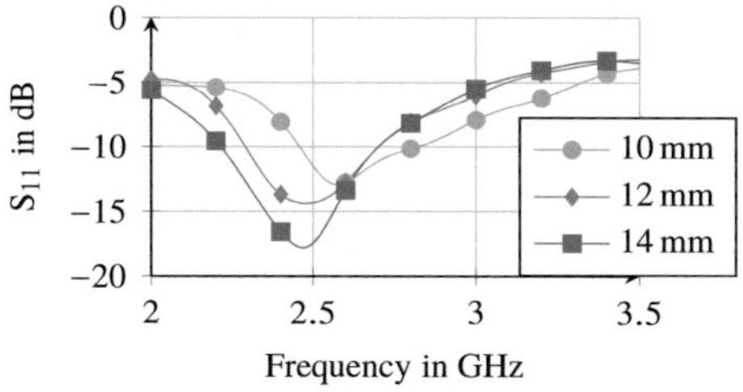


(a) Sweep of $l_h$ for $l_v$ fixed to 10 mm

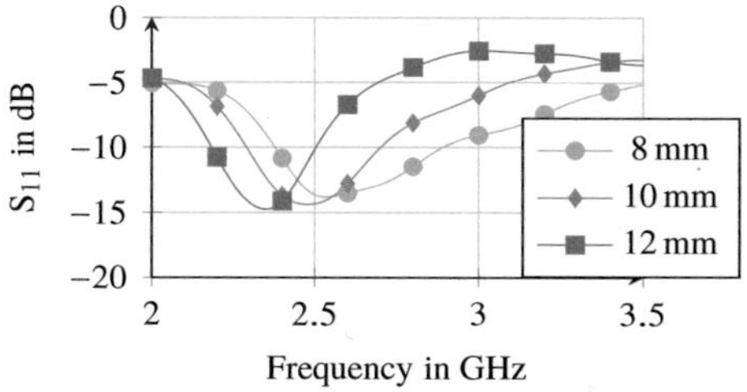


(b) Sweep of $l_v$ for $l_h$ fixed to 12 mm

Figure 4.24: Simulation results presenting the effect of the length of the vertical and the horizontal bar on the antenna's reflection coefficient.

gain value at 2.5 GHz is 5.9 dBi, 5.3 dBi and 4.1 dBi respectively (see table 4.7). Therefore, a compromise between antenna miniaturization and performance is met and $l_h$ and $l_v$ are set to 12 mm and 10 mm respectively.

Three reconfigurable patterns can be realized with the presented antenna. Switching the right p-i-n diode to OFF state and the left one to ON state activates state 1. In this case the left radiator is fed and the right one acts as a reflector, resulting in a beam pointing in the negative $x$ direction (see Fig. 4.25). If both p-i-n diodes are ON, the antenna radiates nearly omnidirectional, however with increased gain along the $z$-axis (state 2). In state 3, the right diode is ON and the left one is OFF, and a beam pointing in the positive $x$ direction is obtained (see Fig. 4.25).

Table 4.7: Simulation results presenting the effect of the length of the vertical and the horizontal bar on the antenna's gain

| Sweep of $l_h$ for $l_v$ fixed to 10 mm | | | |
|---|---|---|---|
| $l_h$ in mm | 10 | 12 | 14 |
| Gain at 2.5 GHz in dBi | 6.6 | 5.8 | 5.4 |
| Sweep of $l_v$ for $l_h$ fixed to 12 mm | | | |
| $l_v$ in mm | 8 | 10 | 12 |
| Gain at 2.5 GHz in dBi | 5.9 | 5.3 | 4.1 |

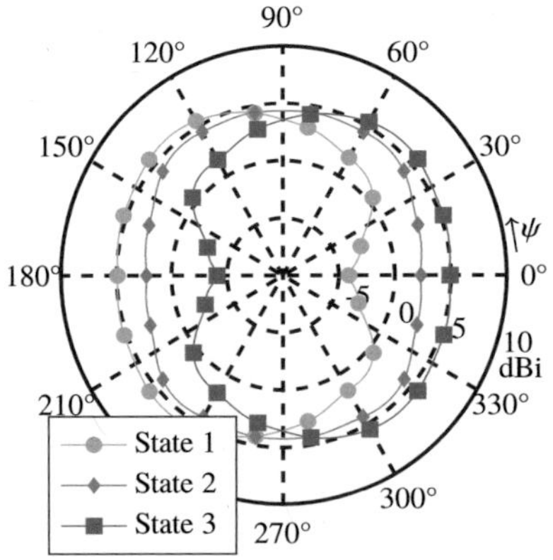


Figure 4.25: Simulated horizontal cuts of the radiation patterns for a single element of the wideband multiple antenna system at 2.5 GHz for elevation angle $\theta$ = 60°.

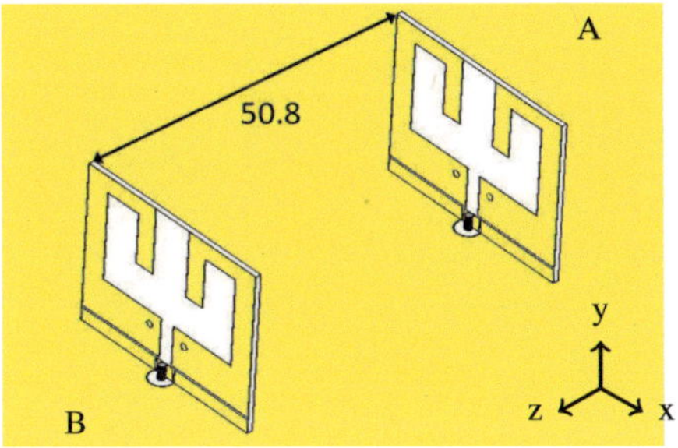


Figure 4.26: Configuration of the wideband MIMO reconfigurable antenna (perspective view). Dimensions are given in mm [21].

### 4.4.2 MIMO Configuration

Two single antennas placed in 50.8 mm ($0.42\lambda_0$) distance along the $z$ axis constitute a multiple antenna system. The antennas are parallel to one another and referred to as antenna A and B (see Fig. 4.26). The distance between the antennas is a compromise between minimal footprint of the MIMO antenna and the isolation between both antennas.

The antennas A and B show good performance in terms of simulated input reflection coefficients and isolation (see Fig. 4.27). Considering the −6 dB bandwidth, the antennas cover the frequency range from 2.1 to 2.87 GHz (31 % fractional bandwidth). The lowest isolation is observed if both antennas are in state 2. In this case substantial part of the E-field radiated by both antennas is directed towards the other antenna what explains lower isolation. Nevertheless, even in this case it is better than 12 dB. Since state 1 and 3 are symmetric about the $z$-axis, only state 1 is presented in the plot.

The simulated patterns of antenna A point towards the rear-left, left and front-left side of the vehicle (see Fig. 4.28a). For antenna B the beams point towards the rear-right, right and front-right side of the vehicle (see Fig. 4.28b). Compared to the single antenna, radiation of the antennas in MIMO system is shifted towards sides of the vehicle. It is caused by the shadowing between the antennas along $z$-axis. The patterns for antenna A are shifted in the negative $z$ direction. In opposite the patterns for antenna B are shifted in the positive $z$ direction. This is especially visible for state 2 of both antennas. The obtained gain varies between

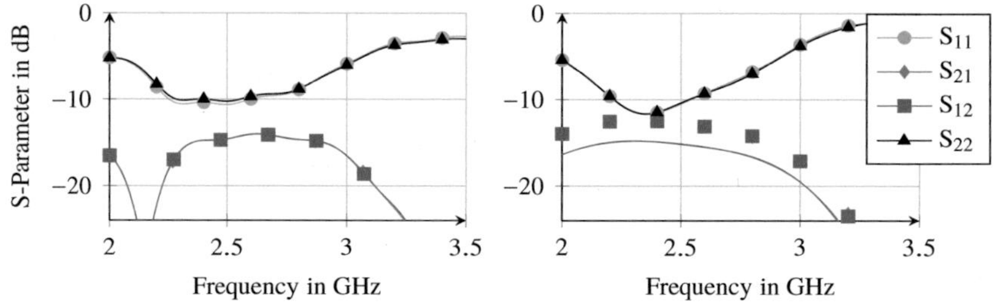


(a) State 1 (b) State 2

Figure 4.27: Simulated S-parameters for the presented wideband MIMO antenna. The antennas A and B are associated to port 1 and 2 respectively.

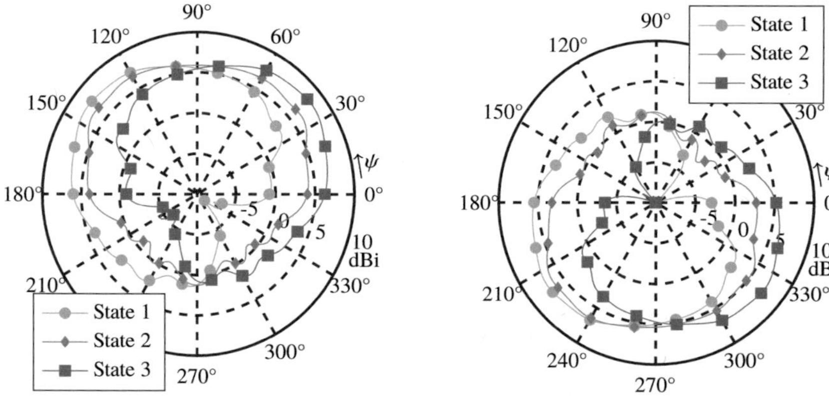


(a) Antenna A (b) Antenna B

Figure 4.28: Simulated horizontal cuts of the radiation patterns for the presented wideband MIMO antenna system at 2.5 GHz for elevation angle $\theta = 60°$.

7.1 dBi for state 2 and 7.9 dBi for states 1 and 3 at 2.5 GHz and is higher by up to 2 dB compared to the single antenna.

In the next step a prototype of the antenna system was fabricated. The antennas are etched on 0.508 mm thick Rogers RT5880 substrate. Once again the NXP BAP64-02 p-i-n diodes were used. The antennas were mounted on a metal plated FR-4 board used as a ground plane (see Fig. 4.29).

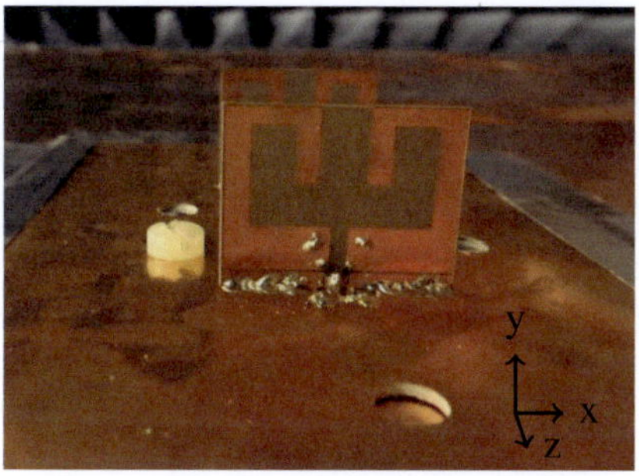


Figure 4.29: Fabricated prototype of the wideband multiple antenna system [21].

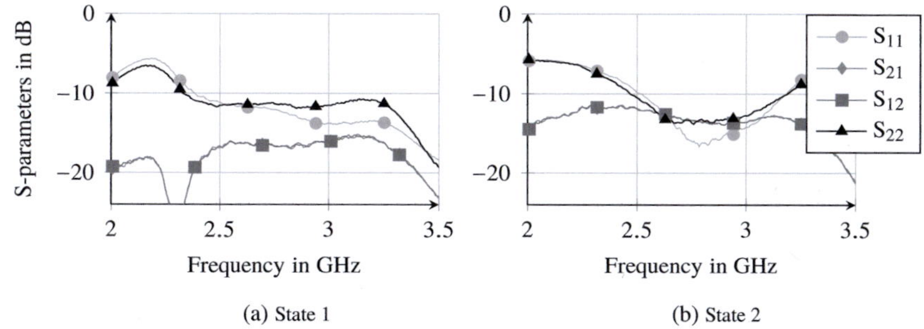


Figure 4.30: Measured S-parameters for the presented wideband MIMO antenna. Both single antennas are in the same reconfigurable state.

The measured S-parameters for states 1 and 2 of both antennas are depicted in Fig. 4.30 (state 3 is not plotted as it is symmetric to state 1). The matching is slightly better than expected from the simulation. Especially at the frequencies above 2.6 GHz matching is better than simulated. The prototype of the antenna system covers the frequency range from 2.1 to 3.4 GHz (−6 dB bandwidth).

Following, gain pattern measurements were performed inside an anechoic chamber. The drill holes visible in the Fig. 4.29 are necessary to mount the antenna on the rotary tower inside the anechoic chamber. During the measurements the structure was mounted on a 500 mm × 500 mm ground plane. The reconfigurable states of the antenna are switched ON by applying 5 V from a DC-power supply to the appropriate p-i-n diodes, while the other diode is connected to

Table 4.8: Radiation pattern parameters of the presented wideband MIMO antenna system.

| State | Gain in dBi | Beam direction $\psi$ | $\text{HPBW}_{\psi}$ |
|---|---|---|---|
| A1 | 7.3 | 135° | 80° |
| A2 | 5.7 | 90° | 165° |
| A3 | 7.1 | 45° | 80° |
| B1 | 7.3 | 225° | 80° |
| B2 | 5.8 | 270° | 165° |
| B3 | 7.2 | 315° | 80° |

0 V to remain in OFF state. The measured patterns are presented in Fig. 4.31. Their form for different states resembles the simulated patterns very well. Also the measured gain is close to the simulated values. For states 1 and 3 of both antennas maximum values between 7.1 dBi and 7.3 dBi are obtained. Compared to the simulated gain of 7.9 dBi for these states, a difference of around 0.6 dB is observed. This difference can be explained by the losses in a bias-T which is used for DC feed of the antenna. A bigger difference is observed for state 1 of both antennas, where instead of 7.1 dBi a measured gain of 5.7 dBi is observed. In this case higher losses might be accounted to higher insertion loss of p-i-n diodes, since two diodes per antenna are used in this state. The most important radiation parameters of this MIMO antenna system are given in table 4.8.

The ECC values for this antenna system, calculated based on equation 4.1 are below 0.001 making the antenna attractive for MIMO applications. Furthermore, the antenna system realizes patterns with high gain, making it a potential candidate for channel capacity enhancement.

## 4.5 Flush-Mounted Antenna

As discussed in chapter 1 vehicle's roof remains the most attractive mounting position for communication antenna systems. The state of the art solution is mounting inside a so-called shark-fin. However, an attractive new idea is to flush-mount the antennas and the accompanying electronics inside cavities in the chassis [TC17,LJC17]. This novel solution offers more space than the shark-fin, nevertheless the antennas should be low-profile to fit inside the cavity. In scope of

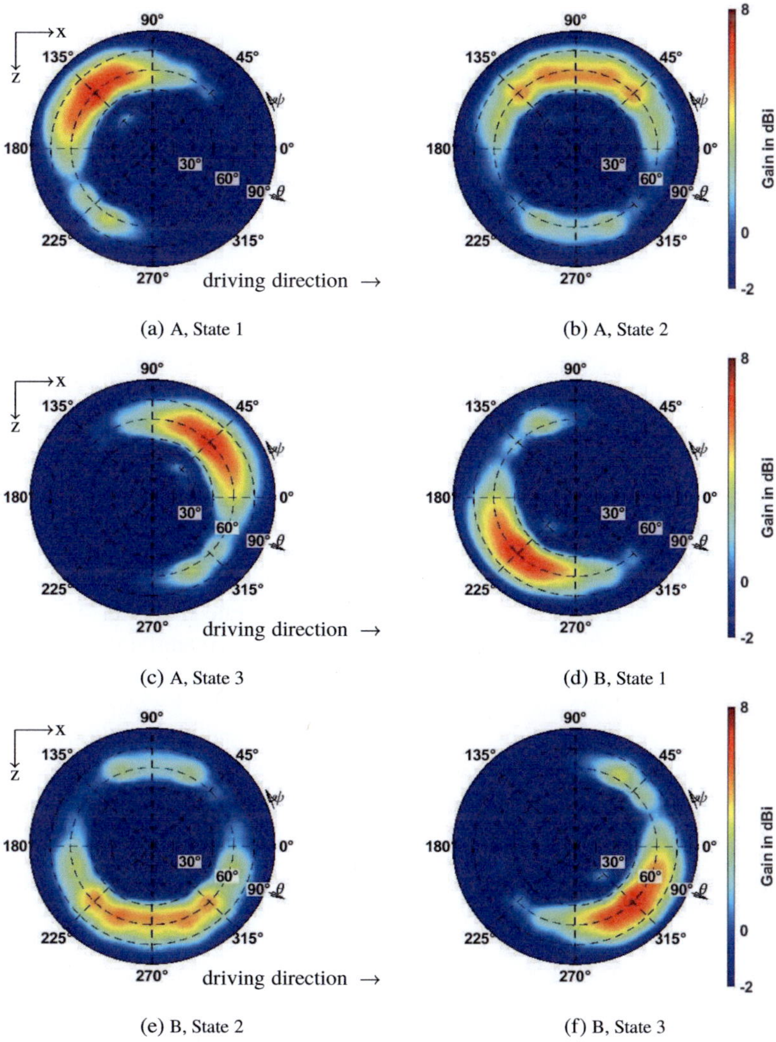


(a) A, State 1

(b) A, State 2

(c) A, State 3

(d) B, State 1

(e) B, State 2

(f) B, State 3

Figure 4.31: Measured radiation patterns of the presented wideband MIMO antenna system at 2.5 GHz.

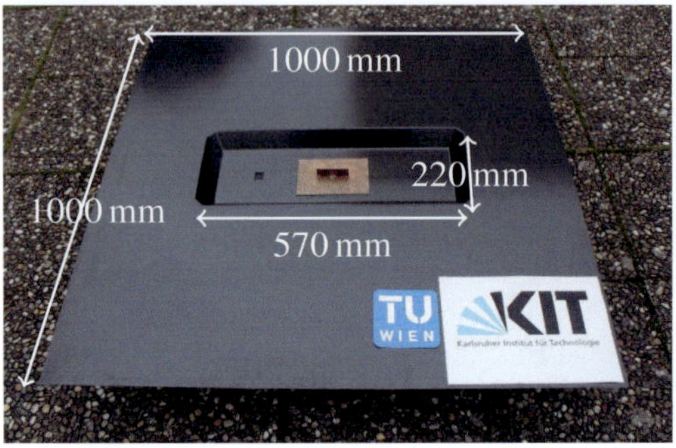


Figure 4.32: CFRP chassis mockup with a reconfigurable antenna inside the cavity.

this work a strong emphasis was put on the miniaturization of antennas, therefore they can successfully be installed within the chassis cavity. This concept was tested in cooperation with colleagues from Technical University of Vienna [6–8].

The group from Technical University of Vienna developed a cavity for vehicular antennas [ALM17]. The cavity is placed in the middle of a $1000 \times 1000\,\mathrm{mm}^2$ carbon fiber reinforced polymer (CFRP) chassis mockup (see Fig. 4.32). The cavity walls are inclined and the overall size is $570 \times 220 \times 40\,\mathrm{mm}^3$. In scope of this cooperation performance of five of the antennas presented in this work was evaluated. The measurements were conducted in the anechoic chamber at KIT. All of the antennas could be successfully operated in their reconfigurable states when flush-mounted. The influence of capacity on the antenna matching is negligible. Yet, some impact on the radiation pattern of the antenna was observed. The differences in the radiation pattern are especially visible on the shorter axis of the cavity. However, the results of this investigation prove that the reconfigurable antennas can be successfully mounted in chassis cavities.

## 4.6 Conclusion

Four pattern reconfigurable multiple antenna systems were presented in this chapter. All of them meet the requirements on bandwidth and radiation pattern directions defined in the introduction of this work. The presented results prove

Table 4.9: Comparison of the four proposed MIMO antenna system

| System | Max Gain in dBi | BW in % | Size in $mm^3$ | States per port | No. of diodes |
|---|---|---|---|---|---|
| 1 | 8.7 | 17 | 66x40x23.4 | 3/3 | 4 |
| 2 | 8 | 15 | 85x45x13.2 | 6/6 | 6 |
| 3 | 8.4 | 8 | 63x65x18 | 4/2 | 8 |
| 4 | 7.3 | 31 | 50.8x30x24 | 3/3 | 4 |

that reconfigurable antennas can be successfully incorporated in MIMO systems increasing the performance and flexibility of antenna systems. A comparison of these antenna systems is given in table 4.9. Among them system 1 has the highest maximum gain, however it is only a simulated value and we have to remember that in case of this system the difference between the simulated and measured gain for the single antenna was the highest. Furthermore, this system requires a relatively big volume, thus it is not considered in the further evaluation. The other three designs will be further evaluated in chapter 5. Among these three antenna systems, system 3 obtains highest maximum gain, however it is the biggest one. Furthermore, it requires 8 p-i-n diodes making the fabrication more expensive. System 2 is characterized by high gain and the highest number of reconfigurable states, thus being the most flexible solution. However, the construction is very complicated and not very robust, making it impracticable for mass production. System 4 has the most simple construction and lowest number of p-i-n diodes used making it robust against fabrication imperfections and applicable in mass production. Moreover, this design is the most compact one and hast the largest fractional bandwidth. Therefore it seems to be the most attractive candidate for automotive applications.

# 5 Evaluation of the designed Antenna Systems

After successfully designing the multiple antenna systems they should be evaluated in order to determine their usefulness in a vehicular application. A choice of proper metric for the antenna system performance was already discussed in section 1.4. In this chapter three of the antenna systems presented in chapter 4 will be first evaluated using the proposed channel based ECC. Next these results will be proved by the channel capacity calculated from RT simulations. Finally, results of a real world test drives in an urban environment located in downtown Karlsruhe (see Fig. 5.1) are presented and discussed.

## 5.1 Channel based ECC calculation

A very good metric for MIMO antenna systems performance is the envelope correlation coefficient (ECC). It is a measure of the correlation between the radiation patterns of the MIMO antenna pairs. ECC calculation requires individual measurements (or simulations) of the complex radiation pattern (magnitude and phase) for each antenna. The less correlated these patterns are, the higher the diversity gain and thus the system's capacity. The ECC ranges from 0 (no correlation, best MIMO gain) to 1 (identical gain and phase patterns, no MIMO gain). This measure can also be used to evaluate the performance of pattern reconfigurable antennas examining the correlation between the different switchable patterns. In standard ECC calculation, a uniform amplitude and phase distribution $P_{\text{Uniform}}(\theta, \psi) \propto 1$ is assumed. This is true for mobile phones, nonetheless in case of vehicular antennas we observe direction selective channel as discussed in section 2.8. Thus, the information about the channel in form of an angular power spectrum (APS) $P(\theta,\psi)$ has to be included. The formula for ECC changes from:

$$\rho_e = \frac{|\iint\limits_{4\pi} \overrightarrow{C_1}(\theta,\psi)\cdot\overrightarrow{C_2}^*(\theta,\psi)d\Omega|^2}{\iint\limits_{4\pi} |\overrightarrow{C_1}(\theta,\psi)|^2 d\Omega \iint\limits_{4\pi} |\overrightarrow{C_2}(\theta,\psi)|^2 d\Omega}, \tag{5.1}$$

to:

$$\rho_{e,APS} = \frac{|\iint\limits_{4\pi} \overrightarrow{C_1}(\theta,\psi)P(\theta,\psi)\cdot\overrightarrow{C_2}^*(\theta,\psi)P(\theta,\psi)d\Omega|^2}{\iint\limits_{4\pi} |\overrightarrow{C_1}(\theta,\psi)P(\theta,\psi)|^2 d\Omega \iint\limits_{4\pi} |\overrightarrow{C_2}(\theta,\psi)P(\theta,\psi)|^2 d\Omega}. \tag{5.2}$$

$\overrightarrow{C}(\theta,\psi)$ denotes here the complex radiation pattern of the antenna.

Since the proper metric is established, the antenna systems can now be evaluated.

### 5.1.1 Evaluation Results

For the sake of evaluation three reconfigurable multiple antenna systems and a multiple antenna system consisting of two omnidirectional antennas (monopoles) were investigated. The omnidirectional antennas were chosen as a benchmark, since this type of antenna is commonly used for vehicular communication. The monopoles are separated by $\lambda_0/2$ at 2.5 GHz. They have nearly omnidirectional patterns in azimuth with slightly higher gain towards front and back of the car (see Fig. 5.2). The difference between the gain of each antenna towards front and rear is due to the shadowing effect appearing between the two monopoles. The maximum simulated gain is for both radiators 5.9 dBi and the −6 dB bandwidth covers the frequency range from 2.2 to 3.7 GHz. The chosen reconfigurable antenna systems are the ones presented in sections: 4.2, 4.3 and 4.4. The antenna systems will be referred to as system I, II and III respectively.

In order to determine channel parameters for ECC analysis, ten routes placed in downtown Karlsruhe (blue region in Fig. 5.1) were used for the RT channel simulation. These routes are different than those used in section 2.8 to provide enough variance between training and test data. Each of the routes consists of

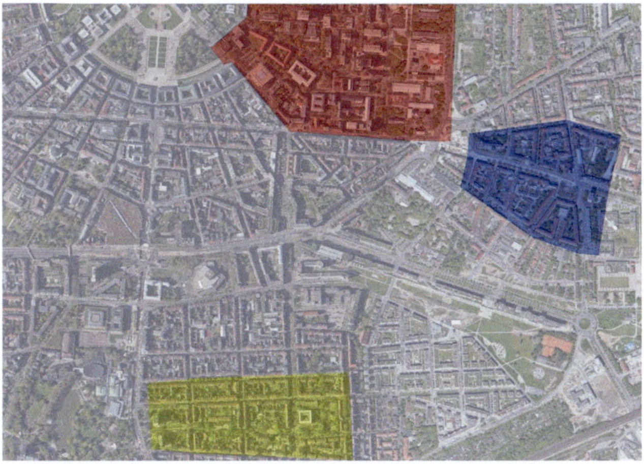

Figure 5.1: Aerial view of Karlsruhe. The different regions considered in this work are marked with colors: red color represents the KIT campus, blue the downtown area in the vicinity of campus and yellow the area where real world and virtual test drives were made. Map: World Imagery, Courtesy of Esri ArcGIS Online.

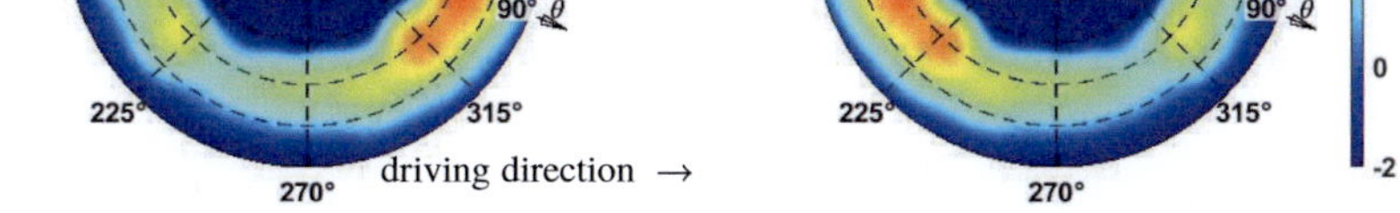


(a) Antenna A (b) Antenna B

Figure 5.2: Simulated radiation patterns of the utilized MIMO antenna system composed of two monopoles at 2.5 GHz.

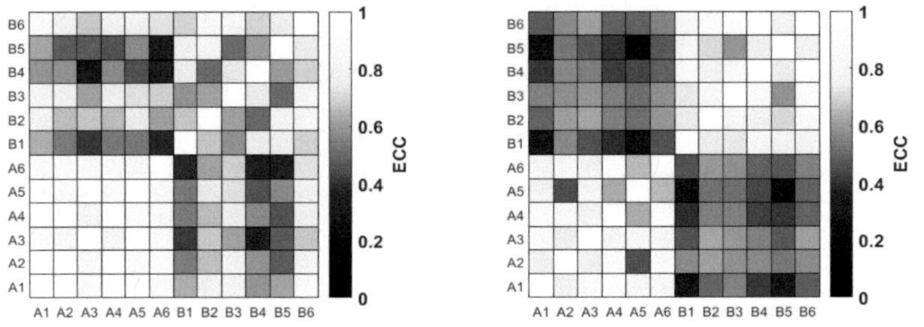


(a) $\delta\psi = 0°$

(b) $\delta\psi = 90°$

Figure 5.3: Channel information based envelope correlation coefficient for different states of antenna system I at 2.5 GHz.

multiple snapshots (time samples). The channel is first averaged within one scenario since very often there are only about ten rays per snapshot, making the array sparse. By producing the mean value over all snapshots a denser occupied array is generated. Next, the APS is multiplied with patterns and the ECC is processed. Finally, the mean value over all scenarios is calculated. The results are presented as a rectangular matrix with ECC values from 0 to 1 represented by the respective color (see Fig. 5.3 to Fig. 5.6). The lower left and the upper right quadrants present the interstate ECC value (decorrelation between reconfigurable patterns), while the lower right and the upper left quadrants represent ECC between different antenna ports. The latter one is crucial for the channel capacity in MIMO systems and the lower the ECC values are, the higher the expected capacity.

Since the automotive channel is direction selective, the orientation of the antenna system in regard to the driving direction is investigated as well. Antenna systems oriented as defined in the original designs in chapter 4 are denoted with $\delta\psi = 0°$. Rotation of the system by 90° is denoted as $\delta\psi = 90°$. The antennas within the system are denoted with A and B, and reconfigurable states are numbered from 1 to 6, depending on the antenna system.

At the first look we can see that all of the proposed reconfigurable antenna systems outperform the conventional system composed of omnidirectional antennas (see Fig. 5.3 to Fig. 5.6). Potentially the highest ECC values are observed for the system of two monopoles (see Fig. 5.6). The next aspect are clear differences in

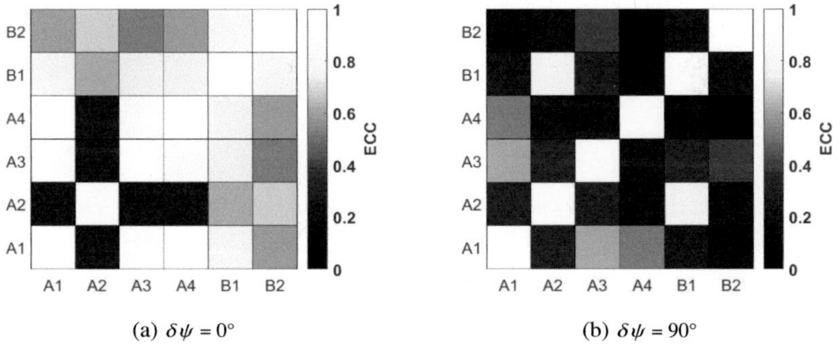


Figure 5.4: Channel information based envelope correlation coefficient for different states of antenna system II at 2.5 GHz.

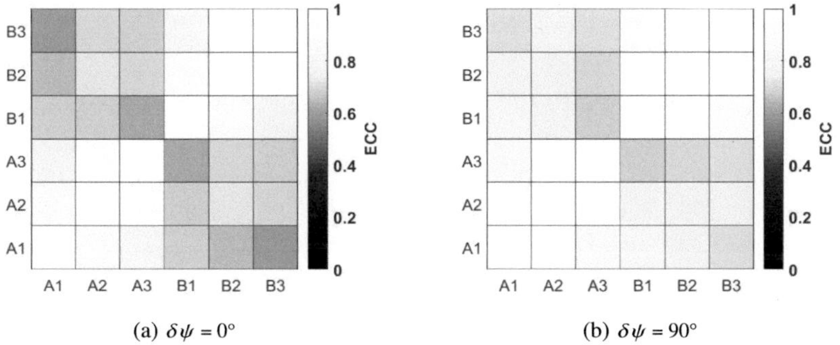


Figure 5.5: Channel information based envelope correlation coefficient for different states of antenna system III at 2.5 GHz.

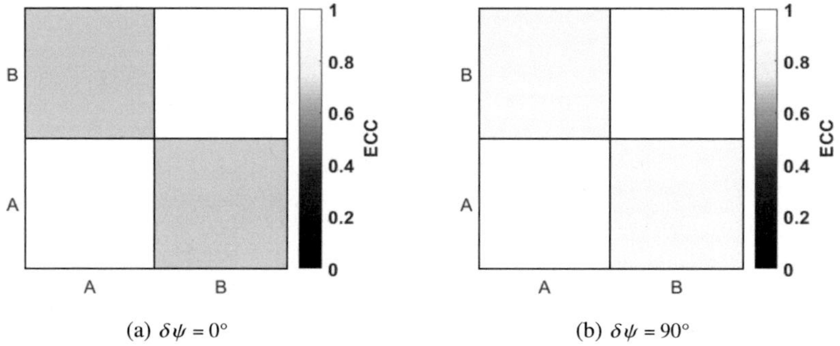


Figure 5.6: Channel information based envelope correlation coefficient for different states of antenna system composed of monopoles at 2.5 GHz.

ECC values between $\delta\psi = 0°$ and $\delta\psi = 90°$ orientation of the systems. We can observe that systems with more directive beams and higher gain, pointing in and against driving direction, bring the most improvement. Similar observation can be made for systems producing beams covering the front-left, front-right, rear-left and rear-right direction e.g. rotating the system I by 90° decreases the ECC values drastically (see Fig.5.3b). This result was to be expected from channel analysis presented in section 2.8.

The best results in terms of ECC are obtained for antenna system II rotated by 90° (see Fig. 5.4b). In this case the values vary between 0.02 (A4 to B2) and 0.91 (A2 to B1). The ECC values between the antenna ports have significantly improved compared to the original setup (0°), while in some cases interstate ECC got worse (e.g. A2 to A1). However, exactly the improvement in ECC between the ports brings higher capacity with itself. To get more insight into these results, it is interesting to analyze the time allotment of the reconfigurable states during the virtual test drive. Looking at the switching statistic, the three most frequent combinations are: A3 with B2, A4 with B2 and A3 with B1 (respectively 29 %, 20 % and 13 % of the time). The improvement in ECC is 0.58 to 0.02, 0.53 to 0.39 and 0.76 to 0.31 for the three combinations respectively. It is worth mentioning, that for the 90° in the states A3 and A4 the beam of antenna A is pointing in either driving direction or against the driving direction with gain exceeding 9.1 dBi (simulated). That explains the capacity improvement in this case.

Also in case of system I, the rotation by 90° brings an improvement (see Fig. 5.3). The ECC values between the ports vary between 0.295 and 0.64. The interstate ECC varies between 0.45 and 0.96. If we have a look at the switching statistics the four most frequent combinations are: A2 with B3, A3 with B2, A2 with B2 and A2 with B1 (respectively 18 %, 16 %, 10 % and 6 % of the time). The improvement in ECC values between $\delta\psi = 0°$ and $\delta\psi = 90°$ varies between 0.04 and 0.25. The beams in states 2 and 3 of antenna A and B point in right-rear, right-front, left-front and left-rear direction. This result once again coincide with the results presented in section 2.8.

For system III the better orientation is $\delta\psi = 0°$ and the ECC values vary between 0.58 and 0.72 (see Fig. 5.5). The interstate ECC varies between 0.79 and 0.98. In this case, the four most frequent combinations are: A1 with B1, A3 with B3, A1 with B3 and A3 with B1 (respectively 33 %, 22 %, 11 % and 9 % of the time). The difference in ECC values between $\delta\psi = 0°$ and $\delta\psi = 90°$ varies between 0.05 and 0.13. The beams in states 1 and 3 point toward the right-rear, right-front, left-front and left-rear side of the vehicle. This correlates well with previous results.

From the presented results we can see that the used method provides the antenna designer with detailed hints considering the performance of antenna system. Furthermore, a good insight into different patterns is available, enabling antenna optimization. Nevertheless, a measure providing a final information about systems performance, the channel capacity, has to be calculated to revise these results.

## 5.2 Channel Capacity Calculation

The channel capacity is a measure of the information rate in a transmission channel. It describes an upper bound and the transmission with an information rate below this limit can possibly achieve an error-free transmission through the channel [PS09]. The capacity was defined by Shannon in 1948 and demonstrates the effect of transmit power, bandwidth, and additive noise on the capacity. Unlike throughput it is independent of the used modulation, making it more appropriate for bench-marking of hardware solutions. The basic formula for the channel capacity is

$$C = \log_2\left(1 + \frac{S}{N}\right), \tag{5.3}$$

where $S$ is transmitted signal power and $N$ is the power of the additive noise.

In case of this work the capacity of MIMO channels is of interest. To simplify the calculation we assume a frequency-nonselective channel known at the receiver. In this case the channel matrix can be expressed as $\boldsymbol{H}$ as explained in section 2.6. The channel capacity between the $M$ transmit antennas and the $N$ receive antennas is then expressed by the equation 2.19. We can simplify this equation to

$$C = \sum_{n=1}^{M} \log_2 \left\{ \left\| \left( \boldsymbol{I}_N + \frac{\mathrm{SNR}}{M} \cdot \boldsymbol{H}\boldsymbol{H}^H \right) \right\| \right\} \tag{5.4}$$

which can be estimated from the channel matrix $\boldsymbol{H}$ and the signal-to-noise ratio (SNR) at the receiver. $\boldsymbol{I}_N$ represents the identity matrix of dimension $N$ and $(\cdot)^H$ denotes the Hermitian transpose. Due to the overall power constraint, the Frobenius norm [PS09] of the channel matrix must satisfy $||\boldsymbol{H}||_{\mathrm{F}}^2 = M \cdot N$.

The channel parameters obtained from the RT simulations were used for the channel capacity calculation. The same ten routes as in section 5.1.1 simulated in the RT were used in this case. A radiation pattern of a typical BS panel antenna with a maximu gain of 17.5 dBi (half power beam width (HPBW) of 63° and 5.2° in azimuth and elevation), similar to [CKC+18], was used in the calculation. On the receiver side (vehicular antenna) patterns of the same reconfigurable antenna systems and monopoles as discussed in section 5.1 are applied. The transmit power is set to 17 W (around 42 dBm [ETSc]). The noise power is calculated from formula:

$$N = kTB, \tag{5.5}$$

where $k$ is the Boltzmann constant, $T$ is the environment temperature in Kelvin and $B$ is the bandwidth of the receiver. The noise figure is assumed to be 2 dB.

In the first simulation step, a MISO channel matrix $\boldsymbol{H}$ is calculated for each receive antenna port and for all reconfigurable states. The transmit and receive antenna patterns and the channel information are used for this operation. Next, the MISO channel matrices are connected in all possible reconfigurable state

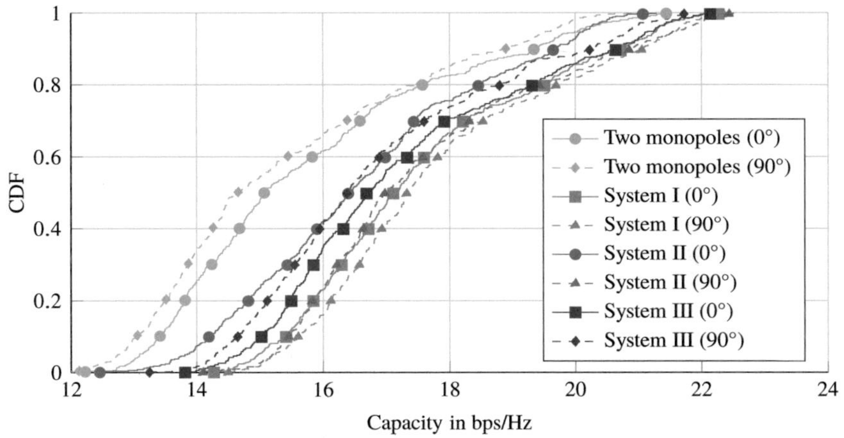


Figure 5.7: CDF plot of the channel capacity calculated from simulated data for the reconfigurable antenna systems and the reference system. The channel is simulated in urban environment at 2.5 GHz.

combinations for antenna A and B, establishing MIMO channel matrices. In the following step, the MIMO channel capacity is calculated. Afterwards, the combination of states providing the highest capacity is selected for each time step. Finally, the average over all scenarios is processed. The results are given in Fig. 5.7.

Looking at the plot we can see that antenna system I rotated by 90° gives the best results in terms of channel capacity. This result is in contrast to the results from section 5.1.1, where antenna system II was characterized by the lowest ECC values. We have to consider that ECC is a measure of correlation and is normalized, thus the information about antenna gain is not included. If we have a look at equation 5.4, we can see that the antenna gain will have an influence on the SNR and therefore effects the channel capacity. Therefore, we can assume that the discrepancy between the results of ECC and channel capacity calculation come from this.

To present and compare the channel capacity for different antenna systems the cumulative distribution function (CDF) is used. The CDF is well known from the probability theory and statistics, and represents the probability that the corresponding capacity $C_k$ will take a value less than or equal to $C$. The CDF can be expressed by formula [PS09]:

$$F(C) = P(C_k \leq C) \tag{5.6}$$

If we for example consider the CDF at 0.1, we will obtain the 10 % outage capacity within analyzed ensemble of all possible channel realizations (on scenario or group of scenarios). It means that the channel capacity will have at least this value for 90 % of the time.

The calculated capacities at CDF 0.1 for antenna system I in $\delta\psi$ = 0° and $\delta\psi$ = 90° orientation are: 15.4 bps/Hz and 15.6 bps/Hz. For system II 14.2 bps/Hz and 15.5 bps/Hz respectively, and for system III 15.1 bps/Hz and 14.7 bps/Hz respectively. Considering an antenna system consisting of two omnidirectional antennas, capacities of 13.4 bps/Hz and 13 bps/Hz for 0° and 90° rotation are obtained. The most significant difference between two antenna system orientations is observed for system II. This is due to beams pointing in either driving direction or against the driving direction with gain exceeding 9.1 dBi for 90° rotation case. Therefore, an important finding is that these two directions are crucial for improvement of the channel capacity in urban scenarios. This issue and switching statistics were already discussed in section 5.1.1.

Although, this work concentrates on capacity enhancement of MIMO antenna systems in urban environment, it has to be considered that the vehicles also drive in different environments like highway or rural surroundings. To prove, if the proposed antenna systems improve the performance in an other environment, the system II and III were tested in terms of channel capacity in a rural scenario. Ten routes placed in a rural environment were simulated in the RT. A radiation pattern of a typical BS panel antenna with a maximu gain of 17.5 dBi as in section 5.2 was used in the calculation. On the receiver side (vehicular antenna) patterns of the same reconfigurable antenna systems and monopoles as discussed previously are applied (this time only system II and III are considered). The transmit power is set to 17 W (around 42 dBm [ETSc]) and the BS height is 30 m. The scenarios are characterized by less obstacles on the signal path compared to the urban case, thus the range is increased.

Similar as in the case of urban scenarios (see section 5.2) also in this case an improvement compared to the reference system can be observed (see Fig. 5.8). The calculated capacities at the CDF 0.1 (10 % outage capacity) for the antenna

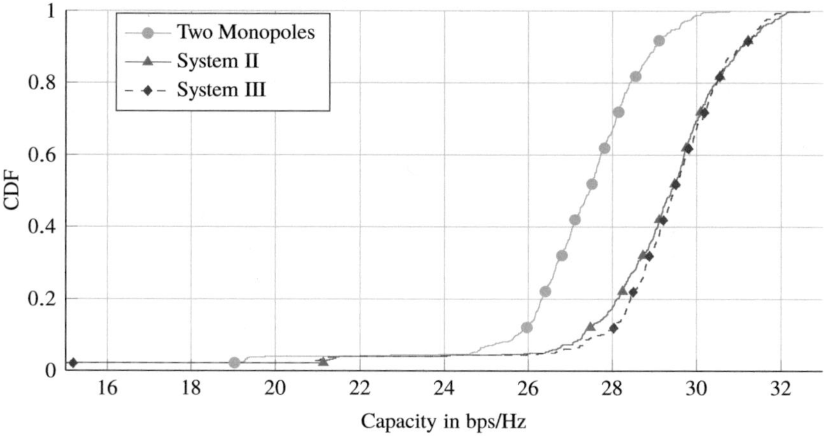


Figure 5.8: CDF plot of the channel capacity calculated from simulated data for the reconfigurable systems and the reference system. The channel is simulated in rural environment at 2.5 GHz.

systems II and III are 27.8 bps/Hz and 27.3 bps/Hz respectively. Considering an antenna system consisting of two omnidirectional antennas, the channel capacity of 24.4 bps/Hz is obtained. Therefore, we observe an improvement by at least 2.9 bps/Hz. In case of the rural scenarios, the LOS component was observed in more snapshots and therefore the antenna patterns pointing towards BS contribute to the channel capacity improvement. The maximum gain values of different reconfigurable patterns of systems II and III exceed the maximum gain value of system consisting of monopoles, which explains these results.

The results show definitely that reconfigurable antenna systems outperform the conventional multiple antenna system. We can observe improvements of 2.2 bps/Hz for the best case and at least 1 bps/Hz at CDF of 0.1 in worst case. It has to be considered that these values are an upper limit for reconfigurable MIMO antenna system, since always the combination of best states was selected in the calculation. To confirm these results by real world measurements, a test drive in urban environment was conducted. As a good candidate antenna system III was selected. The selection is based on good results of channel capacity simulation, compact size and robust construction. The results of the test drive are presented in the following section.

## 5.3 Test Drive

Although the channel simulations using the RT tool give very precise results close to the reality, a measurement in a real wireless channel is substantial for characterization of the vehicular MIMO channel. Some examples of such test procedure are known from the litarature [ETKM13, OOA12], however in those cases only standard antennas were tested. The reconfigurable antenna systems were tested by the channel measurements in [QGWL12, ZAH14], yet only in a laboratory environment. This section presents a test drive in an urban environment including reconfigurable MIMO antennas [3].

### 5.3.1 Measurement Setup

The chosen antenna system was evaluated in downtown Karlsruhe which is a typical urban environment with sparse vegetation, many parked cars and dense traffic flow. This area resembles the one used for the channel simulation in the previous section. However, it is placed in a different part of the city due to better possibility of conducting the measurement campaign. The $M = 2$ transmitters (Tx) were placed at the $10^{\text{th}}$ floor of a hotel as shown in Figure 5.12, a building with the most prominent height in the area. Thus, a realistic placement of a base station at around 30 m height could be mimicked. This is a typical height of BS for the urban macro cells in LTE [ETSc].

The test route is 1.8 km long and includes streets on the east-west axis as well as the north-south axis. Hence, different orientations of the test vehicle towards the transmitter were included. A van type vehicle, available at the institute, was used for the measurements, and the $N = 2$ receive (Rx) antennas were placed on a metallic plate placed in the opening of the sunroof (see Fig. 5.10). The Tx antennas were placed outside the hotel room on the window sill (see Fig. 5.9). On the transmitter side two patch antennas are used. The matching is better than −10 dB at 2.49 GHz and. The maximal gain of 5 dBi, and HPBW of 78° and 88° in azimuth and elevation are measured for both of them.

The monopoles used for the measurement were placed in a distance of 50.8 mm (0.42 $\lambda_0$) along the driving direction of the vehicle. The measured matching is better than −10 dB at 2.49 GHz for both antennas. The front antenna A shows maximum gain of 5.8 dBi and 4.3 dBi towards front and rear of the

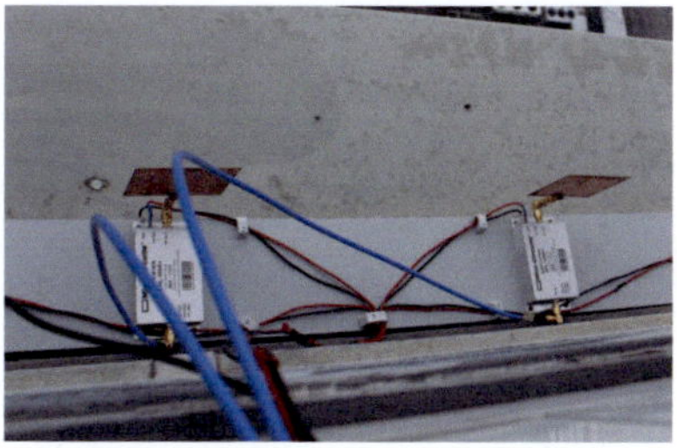

Figure 5.9: Transmitter side of the measurement setup: Tx antennas with amplifiers on the window sill.

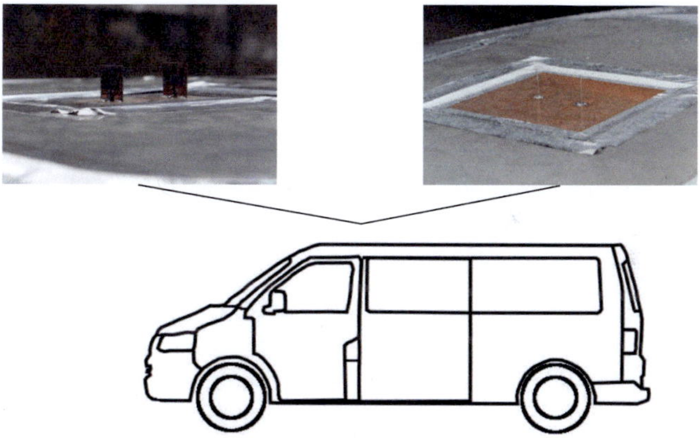

Figure 5.10: Receiver side of the measurement setup: two reconfigurable antennas, and two monopoles placed on the car-roof.

vehicle respectively. For antenna B the radiation pattern is mirrored compared to antenna A. The patterns of the monopoles are not perfectly omnidirectional due to mutual shadowing. The reconfigurable antenna system used here is the same as presented in section 4.4.

The block diagram for the measurement system is shown in figure 5.11. The RF signal processing as well as the digital-to-analogue and analogue-to-digital conversion of the signal is performed by software defined radios (SDRs). As SDR platform a Universal Software Radio Peripheral (USRP) of type X310 from Ettus Research is used. The baseband signal processing is performed on

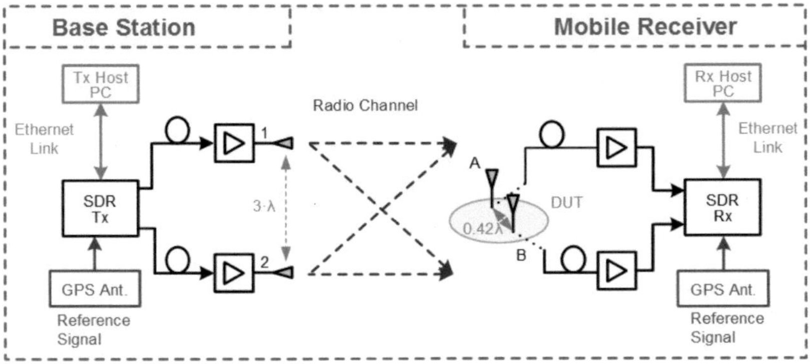


Figure 5.11: Block diagram of the $2 \times 2$ MIMO measurement system.

host PCs, which are connected to the SDRs via an ethernet link. Furthermore, additional power amplifiers (ZRL-3500+ from MiniCircuits) at the Tx and low noise amplifiers (ZX60-33LN-S+ from Mini Circuits) at the Rx are used to improve the link budget. The patch antennas at the Tx are attached directly to the amplifiers and connected to the SDR via cables of equal length. At the Rx side the antennas, LNAs and the SDR are connected as well via cables with the same length. The SDRs at the Tx and Rx are locked to a GPS reference clock to avoid carrier frequency offsets. Moreover, this enables the recording of the receiver location during measurements. To estimate the channel between the Tx and Rx training symbols are used. The estimation is performed assuming a narrowband quasistatic MIMO channel by a least squares algorithm [5]:

$$\boldsymbol{H} = \vec{y} \cdot \vec{x}^{\dagger} \cdot \left(\vec{x} \cdot \vec{x}^{\dagger}\right)^{-1} \tag{5.7}$$

where $\vec{x} = \mathbb{C}^{M \times 1}$ represents the transmitted training signal vector and

$$\vec{y} = \vec{x} \cdot \boldsymbol{H} + \vec{n} \tag{5.8}$$

represents the received signal vector of size $\vec{y} = \mathbb{C}^{N \times 1}$ including an additive white Gaussian noise vector $\vec{n} = \mathbb{C}^{N \times 1}$, which entries follow an independent and identical distribution $\mathbb{C}\mathcal{N}(0, \sigma_n^2)$. The complex conjugate transpose of the signal vector is denoted with $(\cdot)^{\dagger}$.

As signal waveform, orthogonal frequency division multiplexing (OFDM) is utilized. This enables a separation of the Tx antennas by using different OFDM subcarriers. The dedicated subcarriers of each Tx antenna are interleaved to increase the robustness against fading. To minimize the processing effort at the Rx side, the identical OFDM symbol is repeated continuously. Thereby, no time synchronization at the receiver is needed and the cyclic prefix can be omitted [5].

The parameters of the measurement system are listed in table. 5.1. The signal bandwidth is set to 1 MHz to guarantee a frequency flat channel. The center frequency $f_c = 2.49\,\text{GHz}$ is selected. The choice was based on placing it within outer channel of 2.4 GHz ISM band to avoid interference and not to require a license.

Table 5.1: Measurement system parameters.

| Parameter | | Value |
|---|---|---|
| MIMO size | $M \times N$ | $2 \times 2$ |
| Center frequency | $f_c$ | 2.49 GHz |
| Transmit power | $P_{\mathrm{Tx}}$ | 25 dBm |
| Tx antenna gain | $G_{\mathrm{Tx}}$ | 5 dBi |
| Tx antenna spacing | $d_{\mathrm{Tx}}$ | $3 \cdot \lambda_0$ |
| Rx antenna A/B gain | $G_{\mathrm{Rx}}$ | 5.8 dBi |
| Rx antenna A/B spacing | $d_{\mathrm{Rx}}$ | $0.42\lambda_0$ |
| Sampling frequency | $f_S$ | 1 MSps |
| OFDM subcarrier spacing | $\Delta f = f_s / N_{\mathrm{FFT}}$ | 7.8125 kHz |
| FFT size | $N_{\mathrm{FFT}}$ | 128 points |
| OFDM symbol duration | $T_{\mathrm{ofdm}}$ | 128 µs |
| Digital modulation scheme | | 8-PSK |

The evaluation of the measurements takes place in the postprocessing. Therefore, the received data is digitized and recorded during the test run. In the postprocessing the performance is then evaluated at each recorded position. The channel capacity [PS09] is used as a performance measure. The capacity can be estimated from the channel matrix $\boldsymbol{H}$ and the signal-to-noise ratio (SNR) as in equation 5.4. To calculate the capacity within the measurements a quasi SNR is estimated at each receiver after equalization. As quasi SNR the averaged modulation error ratio (MER) over all received channels is used. Furthermore, the channel matrix is averaged over 100 OFDM symbols around the time stamp of the recorded location points.

### 5.3.2 Measurements Results

During the measurements the test route was repeated four times: once with monopoles as Rx antennas (reference) and three times with the presented reconfigurable multiple antenna system on the Rx side. During each one of the three test drives a different reconfigurable state was active at the two antennas. This solution enables switching to the best state in post-processing and at the same time the information about performance of each of the reconfigurable states is available for further analysis. Each of the measurements comprises of 431 Rx locations. Slight differences in the GPS data for each measurement were corrected, so that they are the same for all four test drives. There were 2-3 missing GPS points for every test drive. These were then interpolated using the previous and subsequent GPS points. The GPS positions from the first test drive were taken as reference points. For the GPS points of the subsequent drives, the channel parameters were extracted and then assigned to the nearest GPS position of the reference drive. Finally, the channel capacity for two monopoles and the used recofigurable antenna system were calculated. In case of the presented system a MISO (mutiple-input single-output) channel matrix $\boldsymbol{H}_{1\times2}$ is calculated at the two antenna ports. Those are afterwards combined to a MIMO channel matrix $\boldsymbol{H}_{2\times2}$. Finally, a combination of the best reconfigurable states from both antenna ports is chosen.

The channel capacity for the reference MIMO antenna and the presented antenna system are plotted on a map (see Fig. 5.12) and as a capacity over distance (see Fig. 5.13). The presented antenna system outperforms the reference antenna in most of the cases (86.6 %). The biggest improvement can be observed on a street placed on east-west axis (distance from 0.5 to 1.05 km). The canyoning effect trapping the MPCs in the street is stronger than for the street on the north-south

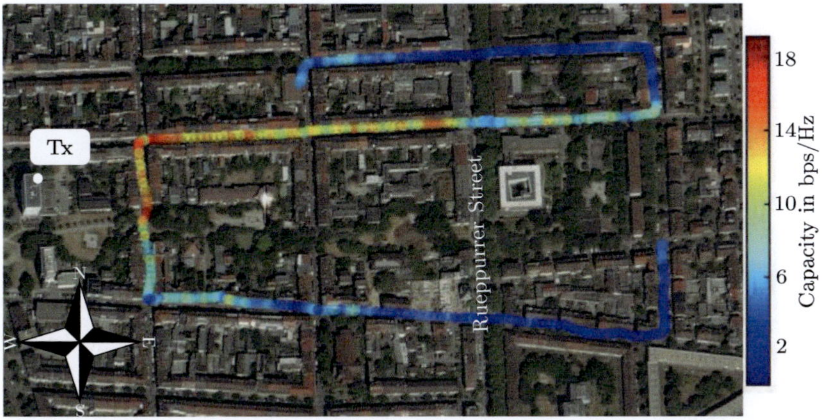


(a) Reconfigurable antenna A & B

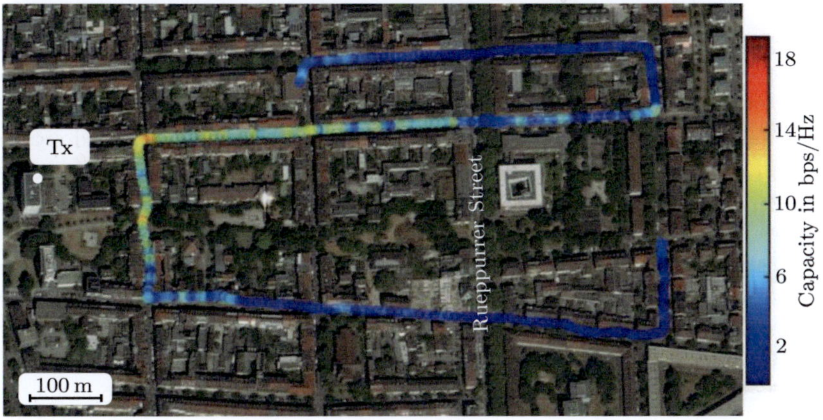


(b) Two monopoles

Figure 5.12: Aerial view of the test route with the channel capacity of a $2 \times 2$ MIMO system calculated from measured data at different receiver positions. Map: World Imagery, Courtesy of Esri ArcGIS Online.

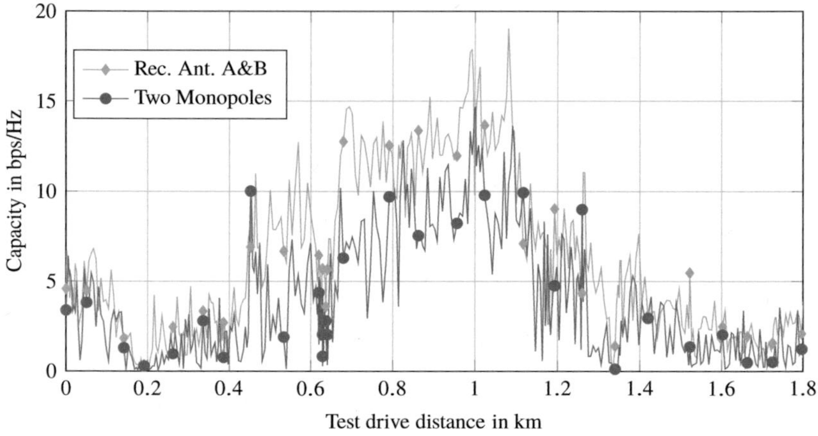


Figure 5.13: MIMO channel capacity calculated from measured data for antennas over the distance of the test drive.

axis (distance from 1.05 to 1.2 km). A drop in channel capacity can be observed whenever Rueppurrer Street (wide street on a nort-south axis) is passed. Two important factors contribute to this: due to the width of this street a gap in a canyon occurs and the MPCs can escape, and the vegetation on the west side of the street blocks the MPCs. However, if these and other cases with low SNR are considered, an improvement in performance of the system when the presented antenna is used can be observed. For the CDF at 0.1 the capacity improves from 0.4 bps/Hz for monopoles to 1.5 bps/Hz (see Fig. 5.14). It is an improvement by a factor 3.75. Considering that only the conventional antenna system was replaced by a reconfigurable one, and the number of front-ends did not change, this is a very good improvement. This result correlates very well with result presented in section 5.2.

Antennas A and B switched the state 134 and 119 times respectively over the test drive distance. Longer periods when one states is used, can be observed in the middle of the route on a long street on the east-west axis (see Fig. 5.15). The states of antenna A and B are switched less often in this case, since due to the structure of the street the MPCs are more stable and their AoA at the antenna system changes less often. In cases with lower SNR the states change more often. This is due to the weaker and less stable MPCs. As it could be expected, the

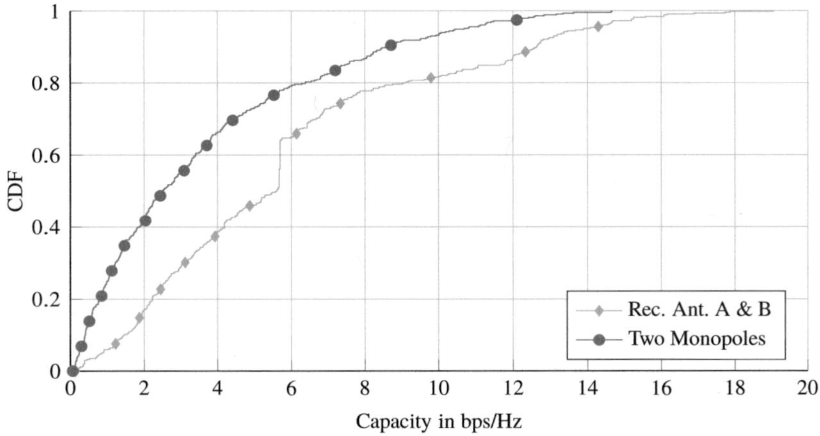


Figure 5.14: CDF plot of the channel capacity calculated from measured data for the reconfigurable multiple antenna system and the two monopoles.

antennas switch their states whenever the vehicle changes the driving direction or passes over a crossroads. Considering the statistics of state usage at antenna A, state 1 is used in 27.6 % of the cases, state 2 in 27.6 % and state 3 in 44.8 %. In case of antenna B states 1, 2 and 3 are used respectively in 22 %, 44 % and 34 % of the cases. This relatively equal distribution proves that determination of radiation direction prior to design leads to an optimized solution.

Another interesting aspect is the analysis of the system's performance in case if instead of optimal switching, the number of state switching cycles is reduced. If we accept that the same state as in previous time step is used whenever it is at least equally good as in the previous step, the number of switching operations is reduced by 14 % and 15 % for antenna A and B respectively. This case is referred to as factor 1 in Fig. 5.16, whereas the case when always the best state is active is referred to as max switch. It can be seen that the CDF at 0.1 almost does not differ between the two cases. However, we can observe a difference of 1.1 bps/Hz at the CDF of 0.5. If we accept that the same state as in previous time step is used whenever it is better than the value from previous step multiplied by a factor 0.8, the number of switching operations is reduced by 31 % and 38 % for antennas A and B. Provided that we repeat the same procedure with a factor 0.6, the number of switching operations is reduced by 48 % and 54 % for antennas

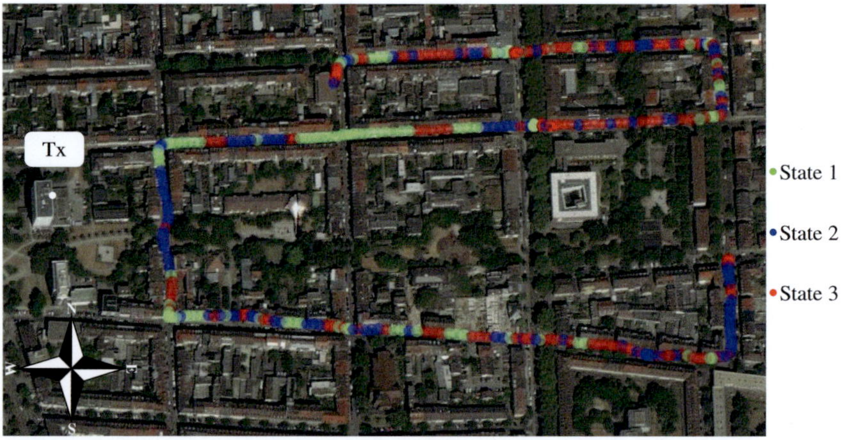


(a) Antenna A

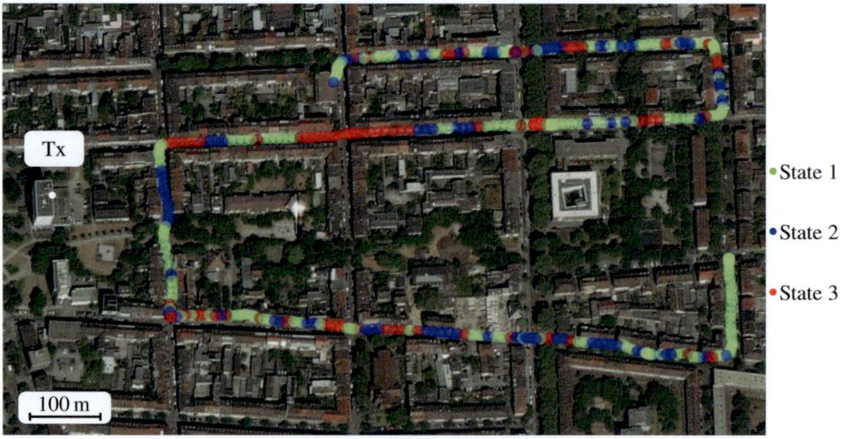


(b) Antenna B

Figure 5.15: Aerial view of the test route with reconfigurable states of antennas A and B at different receiver positions. Map: World Imagery, Courtesy of Esri ArcGIS Online.

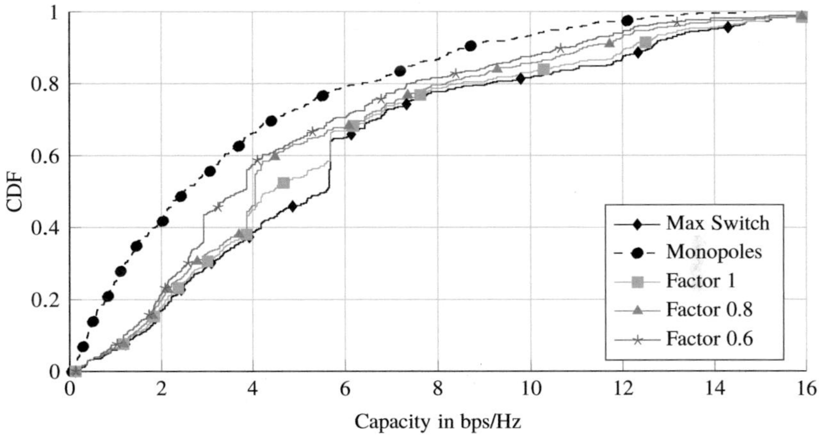


Figure 5.16: CDF plot of the channel capacity calculated from measured data for reduced number of switching operations.

A and B. For the CDF at 0.1 a degradation by maximally 0.3 bps/Hz can be observed, while it is maximally 1.8 bps/Hz for the CDF at 0.5. Nonetheless, in all of the cases the reconfigurable multiple antenna system outperforms the conventional multiple antenna system. Further analysis of the measurement results is presented in appendix A.

For a comparison, the district presented in figure 5.12 was reconstructed in the RT tool. After the simulation, the channel capacity was calculated using the pattern of the patch antenna at Tx and of the test antenna systems at Rx. The Tx power was set to 25 dBm, as in the measurement. The results are plotted in Fig. 5.17. Although also in this case reconfigurable antenna system improves the system's performance, the improvement is lower than in the measurement. The difference origins from the simplified model of the environment used in the RT tool, thus many MPCs existing in the reality do not appear in the simulation. Nevertheless, the course of the curve corresponds well with the measurement.

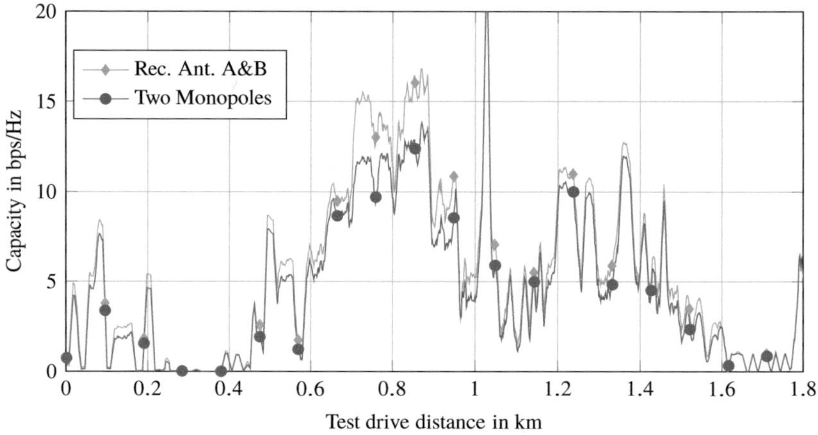


Figure 5.17: MIMO channel capacity calculated from simulated scenario data for antennas over the distance.

## 5.4 Conclusion

In this chapter, it was presented that a channel based ECC is a helpful method for evaluation of reconfigurable antenna systems. The method can successfully support the antenna design process and provide suggestion for optimization. Nonetheless, since the information about the antenna gain gets lost in the normalization process, the ECC is not conclusive. The final design should be evaluated based on channel capacity calculation from both simulated and measured channel values. The results of the capacity calculation confirmed that the use of pattern reconfigurable antenna systems leads to significant performance improvement compared to conventional multiple antenna systems. Results based on channel simulation show an improvement of 2.2 bps/Hz for the best case and at least 1 bps/Hz at the CDF of 0.1 in worst case. The test drive in downtown Karlsruhe proved that tested antenna sytem increases the channel capacity by 1.1 bps/Hz compared to conventional antenna system. This means an improvement by at least by 1.1 bps/Hz in 90 % of the cases. Furthermore, it is observed that antennas producing only one of the six predifined directions at the time bring better improvement. This result is in contradiction to results of the antenna synthesis presented in [5], where two directions are covered simultaneously.

It has to be considered that the results of the antenna synthesis might have some artifacts resulting from statistics (number of examined scenarios). Thus, due to the analysis of different scenarios, the directions from single snapshots get clustered generating patterns covering multiple directions. The results of this work show that it is not necessarily required. Furthermore, generating reconfigurable beams covering one direction at the time enables the antenna designer to increase the maximal gain.

# 6 Conclusion and Outlook

We can observe the growing trend towards connectivity of all electronic devices, enabling the users to communicate with them remotely and for the devices themselves to exchange data. Apart from the permanent connectivity users demand high data rates and higher reliability of the systems. Therefore, new antenna solutions, and especially MIMO antenna systems are being commonly applied. Nevertheless, when it comes to vehicular communication this field still has to be explored and antenna solutions need to be optimized. The results of this work show that by utilizing suitable antenna design strategy, a significant improvement of the channel capacity can be achieved. Furthermore, a design methodology based on simulation of urban vehicular channel is proposed. This methodology includes evaluation techniques like calculation of the channel based ECC and of the channel capacity.

An important part of the presented design strategy is the determination of antenna patterns required in urban environment. This was done based on channel simulations using a ray-tracer tool. The ray-tracer tool and scenarios used for the simulations were discussed in detail. The radiation patterns were established by analyzing the angular power spectrum of the simulated channels. The outcome of this work shows that optimized patterns for urban scenarios should cover the front, front-left and front-right direction referred to the vehicle's driving direction, as well as rear, rear-left and rear-right. Furthermore, the results show that the directions should rather be covered independently than simultaneously. Due to observation done in chapter 5 antenna systems generating beams with high gain in one direction, provide more channel capacity.

Different techniques of pattern reconfiguration were investigated. The three examined techniques are:

- element switching
- parasitic arrays

- phase switching,

before designing the final reconfigurable multiple antenna system for channel capacity enhancement. Among them the phase switching between elements offers the least flexibility in terms of beam direction and lowest gain. Nonetheless, it offers the possibility to generate beams perpendicular to the antenna's substrate, resulting in a small footprint of the design. The other two techniques offer comparable flexibility in terms of beam direction. This however depends strongly on the number of elements and thus on the complexity and size of the antenna. Still, given the required beam directions, compact designs can be achieved using both techniques. Additionally p-i-n diodes and MEMS switches were compared in a practical design. The results of this study show that, if switching elements currently available on the market are used, in case of reconfigurable antennas, use of MEMS switches does not result in better performance, while DC feeding gets more complicated. Therefore, it can be assumed that p-i-n diodes are the proper switching element for automotive communication applications.

The experience collected in the study of single pattern reconfigurable antennas was eventually applied in the design of multiple antenna systems. Four designs are presented and discussed in detail. All of the designs realize the patterns determined in the first part of this work. The antenna systems differ however in their number of reconfigurable states, used switching technique, gain and size. To name two examples: the multiple ESPAR antenna system is a low-profile solution offering high gain and flexibility in terms of beam direction. Thus, it seems to be an attractive solution for automotive applications. However, it suffers on low matching, a big footprint and a complex construction. The wideband multiple antenna system presented in this work fulfills all of the defined requirements on an automotive communication antenna system. Furthermore, with its $50.8 \times 30 \times 24\,\mathrm{mm}^3$ overall size, it is very compact and smaller by 28 % than the ESPAR antenna system.

To prove the potential performance improvement of reconfigurable antenna systems, an evaluation procedure was proposed. Three evaluation methods, being part of the proposed antenna design procedure were used to test some of the presented antenna systems and compare them to a conventional antenna system. A key component in the evaluation process are channel properties combing the properties of antennas and the properties of the propagation medium. The channel based ECC is proposed as a method to support the antenna design process and provide suggestions for its optimization. The effectiveness of this method was verified on an example of selected vehicular antenna systems. The results of this verification show that this measure can be used to optimize an antenna system. However, it is not a sufficient measure to compare various antenna systems, especially if their maximum gain differs. The final verification is given by the channel capacity calculation from both simulated and measured channel values. The results confirmed that use of pattern reconfigurable antenna system leads to significant performance improvement compared to conventional multiple antenna systems. Results based on channel simulation show an improvement between 1 bps/Hz and 2.2 bps/Hz at the CDF of 0.1 compared to a conventional omnidirectional antenna system. To measure real world performance one reconfigurable system was selected and used during a test drive in downtown Karlsruhe. A channel sounder system based on software defined radios was utilized for this purpose. The measurement results proved that the selected reconfigurable antenna system increases the channel capacity by at least 1.1 bps/Hz (from 0.4 bps/Hz for monopoles to 1.5 bps/Hz) in 90 % of the cases compared to a conventional system. It is an improvement by a factor 3.75. Considering that only the conventional antenna system was replaced by the reconfigurable one, and the number of front-ends did not change, this is a very good improvement.

Results of this work not only confirm that the use of pattern reconfigurable multiple antenna systems leads to channel capacity enhancement in automotive scenarios. An even more significant outcome is that, if the environment of the antenna and the propagation channel can be well defined, using a proper design and test procedure based on channel knowledge is inevitable for a proper antenna system optimization. The methodology proposed in this work can successfully be applied for other antenna systems on mobile platforms moving along predefined tracks like e.g. robots in an industrial environment.

# A Analysis of Measured Symbol Bins from three different GPS Positions

As mentioned in chapter 5, 431 Rx locations were acquired and the channel matrix per each recorded location points is averaged over 50 OFDM symbols around the time stamp. It means the channel is averaged over 12.8 ms. An important question is how does the channel change within this time and is the information about the best reconfigurable state not falsified by this averaging. In order to investigate this, three points out of 431 Rx locations are chosen. Fist position is at the distance of 60 m. In this case moderate channel capacity is observed. Second position is at the distance of 190 m. In this case channel capacity close to 0 bps/Hz is observed. Finally, a position at the distance of 750 m with high channel capacity is chosen. For the first position we can observe that one of the states is better than the other two (see Fig. A.1). We can observe similar result for the third position as well (see Fig. A.3). Thus, it is confirmed that there are substantial differences in performance of the reconfigurable states and switching works well and reliable. The situation is different in case of position with low SNR. Here we cannot see clearly which of the states is the best (see Fig. A.2). Potentially state 1 offers the highest SNR for antenna B, while for antenna A state 1 is the best until around 7 ms and state 3 afterwards (see Fig. A.2a). Although it is not clear which of the states is the best and thus the switching is not reliable in this case, we have to remember that the SNR is very low. Even if the state switching does not work fully reliably in this case, the influence on the general results is minor. Therefore, these analysis proves that results presented in chapter 5 and the estimated number of switching cycles are correct.

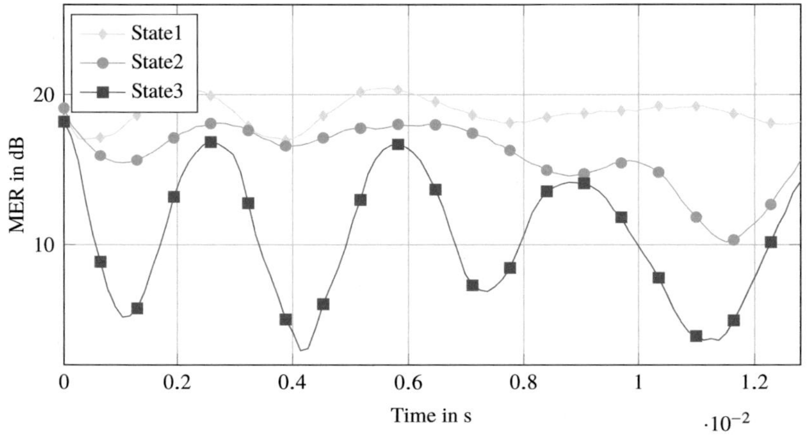


(a) Antenna A

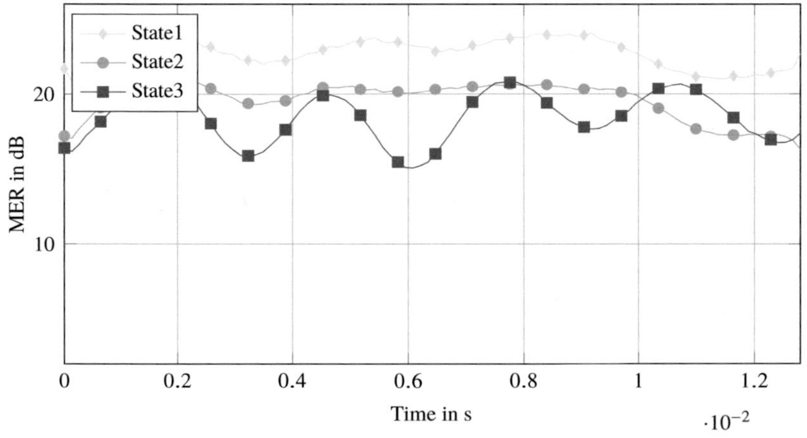


(b) Antenna B

Figure A.1: MER within one bin of 100 OFDM symbols for three states of antennas A and B over time for measurement at 60 m.

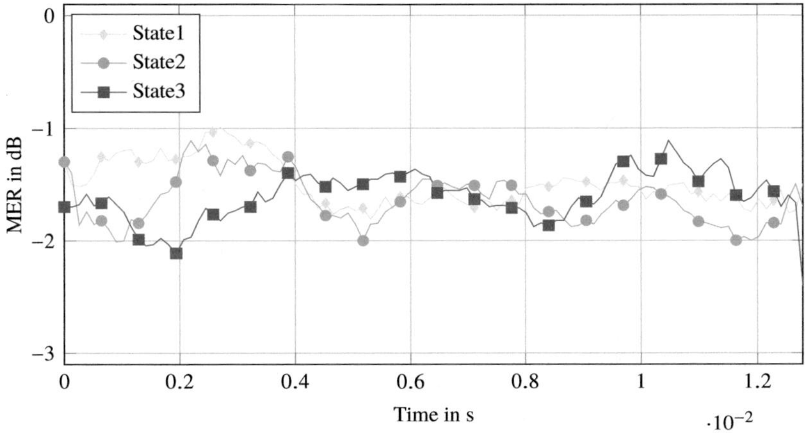


(a) Antenna A

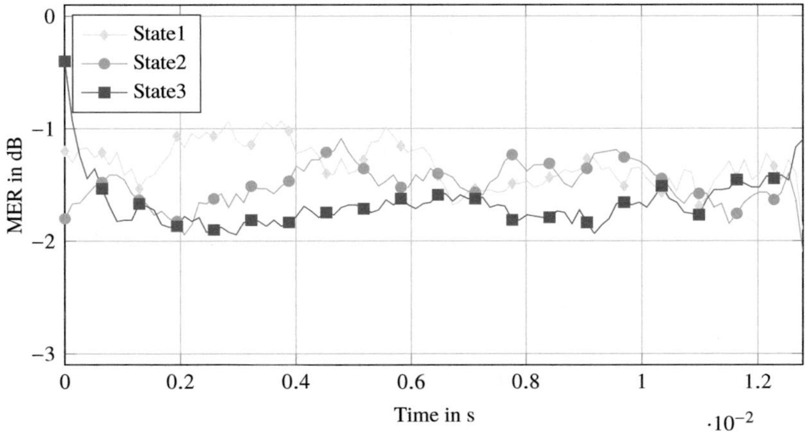


(b) Antenna B

Figure A.2: MER within one bin of 100 OFDM symbols for three states of antennas A and B over time for measurement at 190 m.

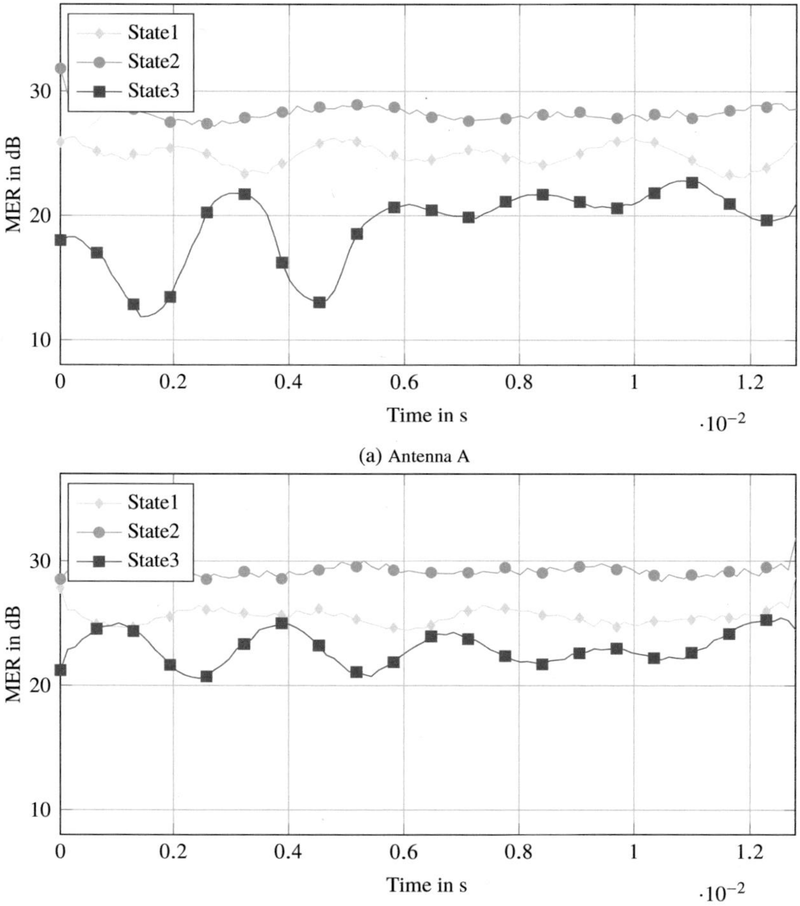


(a) Antenna A



(b) Antenna B

Figure A.3: MER within one bin of 100 OFDM symbols for three states of antennas A and B over time for measurement at 750 m.

# Bibliography


[AA16] M. S. Alam and A. Abbosh. Planar pattern reconfigurable antenna with eight switchable beams for WiMax and WLAN applications. *IET Microwaves, Antennas Propagation*, 10(10):1030–1035, 2016.

[AIG90] McNamara D. A., Pistorius C. W. I., and Malherbe J. A. G. *Introduction to the Uniform Geometrical Theory of Diffraction*. Artech House, Boston, MA, 1990.

[ALM17] G. Artner, R. Langwieser, and C. F. Mecklenbräuker. Concealed CFRP Vehicle Chassis Antenna Cavity. *IEEE Antennas and Wireless Propagation Letters*, 16:1415–1418, 2017.

[Alt] Altair. WinProp. https://altairhyperworks.com/product /FEKO/WinProp-Propagation-Modeling. (visited: 2018-11-08).

[Bal05] Constantine A. Balanis. *Antenna Theory : analysis and design*. Wiley-Interscience, Hoboken, N.J., 3. ed. edition, 2005.

[Ber07] Jennifer T. Bernhard. *Reconfigurable Antennas*. Synthesis Lectures on Antennas. Morgan & Claypool Publishers, 2007.

[Big07] Ezio Biglieri, editor. *MIMO wireless communications*. Cambridge Univ. Press, Cambridge, 1. publ. edition, 2007.

[BRC03] S. Blanch, J. Romeu, and I. Corbella. Exact representation of antenna system diversity performance from input parameter description. *Electronics Letters*, 39(9):705–707, May 2003.

[Bre59] D. G. Brennan. Linear Diversity Combining Techniques. *Proceedings of the IRE*, 47(6):1075–1102, June 1959.

[CCMR07] C. Chiu, C. Cheng, R. D. Murch, and C. R. Rowell. Reduction of Mutual Coupling Between Closely-Packed Antenna Elements. *IEEE Transactions on Antennas and Propagation*, 55(6):1732–1738, June 2007.

[CKC+18] J. Cao, D. Kong, M. Charitos, D. Berkovskyy, A. A. Goulianos, T. Mizutani, F. Tila, G. Hilton, A. Doufexi, and A. Nix. Design and Verification of a Virtual Drive Test Methodology for Vehicular LTE-A Applications. *IEEE Transactions on Vehicular Technology*, 67(5):3791–3799, May 2018.

[COII04] C. Caloz, H. Okabe, T. Iwai, and T. Itoh. A simple and accurate model for microstrip structures with slotted ground plane. *IEEE Microwave and Wireless Components Letters*, 14(4):133–135, April 2004.

[DBG+10] K. Dar, M. Bakhouya, J. Gaber, M. Wack, and P. Lorenz. Wireless communication technologies for ITS applications [Topics in Automotive Networking]. *IEEE Communications Magazine*, 48(5):156–162, May 2010.

[DGM16] A. Dey, R. Guldiken, and G. Mumcu. Microfluidically Reconfigured Wideband Frequency-Tunable Liquid-Metal Monopole Antenna. *IEEE Transactions on Antennas and Propagation*, 64(6):2572–2576, June 2016.

[EPKM14] L. Ekiz, A. Posselt, O. Klemp, and C. F. Mecklenbrauker. System Level Assessment of Vehicular MIMO Antennas in 4G LTE Live Networks. In *2014 IEEE 80th Vehicular Technology Conference (VTC2014-Fall)*, pages 1–5, Sept 2014.

[ETKM13] L. Ekiz, A. Thiel, O. Klemp, and C. F. Mecklenbräuker. MIMO performance evaluation of automotive qualified LTE antennas. In *2013 7th European Conference on Antennas and Propagation (EuCAP)*, pages 1412–1416, April 2013.

[ETSa] ETSI. 4G. https://www.etsi.org/technologies/mobile/4g. (visited: 2019-05-27).

[ETSb] ETSI. 5G. https://www.etsi.org/technologies/mobile/5g. (visited: 2019-05-27).

[ETSc] ETSI. ETSI TR 136 942 V9.1.0. https://www.etsi.org/deliver/etsi_tr/136900_136999/136942/09.01.00_60/tr_136942v090100p.pdf. (visited: 2019-05-31).

[FDAZ+18] M. Fallgren, M. Dillinger, J. Alonso-Zarat, M. Boban, T. Abbas, K. Manolakis, T. Mahmoodi, T. Svensson, A. Laya, and R. Vilalta. Fifth-Generation Technologies for the Connected Car: Capable

Systems for Vehicle-to-Anything Communications. *IEEE Vehicular Technology Magazine*, 13(3):28–38, 2018.

[GL14] I. Goncharova and S. Lindenmeier. A high-efficient 3-D Nefer-antenna for LTE communication on a car. In *The 8th European Conference on Antennas and Propagation (EuCAP 2014)*, pages 3273–3277, April 2014.

[GL15] I. Goncharova and S. Lindenmeier. A high efficient automotive roof-antenna concept for LTE, DAB-L, GNSS and SDARS with low mutual coupling. In *2015 9th European Conference on Antennas and Propagation (EuCAP)*, pages 1–5, April 2015.

[GLG+16] J. S. Gibson, X. Liu, S. V. Georgakopoulos, J. J. Wie, T. H. Ware, and T. J. White. Reconfigurable Antennas Based on Self-Morphing Liquid Crystalline Elastomers. *IEEE Access*, 4:2340–2348, 2016.

[GLW+14] T. Guo, W. Leng, A. Wang, J. Li, and Q. Zhang. A Novel Planar Parasitic Array Antenna With Frequency- and Pattern-Reconfigurable Characteristics. *IEEE Antennas and Wireless Propagation Letters*, 13:1569–1572, 2014.

[GO00] K. Gyoda and T. Ohira. Design of electronically steerable passive array radiator (ESPAR) antennas. In *IEEE Antennas and Propagation Society International Symposium. Transmitting Waves of Progress to the Next Millennium. 2000 Digest. Held in conjunction with: USNC/URSI National Radio Science Meeting (C*, volume 2, pages 922–925 vol.2, July 2000.

[Gup96] Kuldip C. Gupta, editor. *Microstrip lines and slotlines*. Artech House, Boston [u.a.], 2. ed. edition, 1996.

[GW98] Norbert Geng and Werner Wiesbeck. *Planungsmethoden für die Mobilkommunikation : Funknetzplanung unter realen physikalischen Ausbreitungsbedingungen*. Information und Kommunikation. Springer, Berlin, 1998.

[HPD17] J. Hu, D. Pan, and F. Dai. Microstrip Patch Array Antenna With Reconfigurable Omnidirectional and Directional Patterns Using Bistable Composite Laminates. *IEEE Antennas and Wireless Propagation Letters*, 16:2485–2488, 2017.

[HSN08] N. Honma, T. Seki, and K. Nishikawa. Compact Planar Four-Sector Antenna Comprising Microstrip Yagi-Uda Arrays in a

Square Configuration. *IEEE Antennas and Wireless Propagation Letters*, 7:596–598, 2008.

[IK15] M. A. Islam and N. C. Karmakar. Real-World Implementation Challenges of a Novel Dual-Polarized Compact Printable Chipless RFID Tag. *IEEE Transactions on Microwave Theory and Techniques*, 63(12):4581–4591, Dec 2015.

[JSJK14] M. Jusoh, T. Sabapathy, M. F. Jamlos, and M. R. Kamarudin. Reconfigurable Four-Parasitic-Elements Patch Antenna for High-Gain Beam Switching Application. *IEEE Antennas and Wireless Propagation Letters*, 13:79–82, 2014.

[Kil15] Per-Simon Kildal. *Foundations of antenna engineering : a unified approach for Line-of-Sight and Multipath*. Artech House, Boston, 1. ed. edition, 2015.

[KL13] H. Kang and S. Lim. Electric and Magnetic Mode-Switchable Dual Antenna for Null Compensation. *IEEE Antennas and Wireless Propagation Letters*, 12:300–303, 2013.

[KO05] H. Kawakami and T. Ohira. Electrically steerable passive array radiator (ESPAR) antennas. *IEEE Antennas and Propagation Magazine*, 47(2):43–50, April 2005.

[Lar17] E. G. Larsson. Massive MIMO for 5G: Overview and the road ahead. In *2017 51st Annual Conference on Information Sciences and Systems (CISS)*, pages 1–1, March 2017.

[LJC17] S. Lee, G. Jeoung, and J. Choi. Three-dimensional-printed tapered cavity-backed flush-mountable wideband antenna for UAV. *Microwave and Optical Technology Letters*, 59(12):2975–2981, 2017.

[Mac] Macom. MA4AGBLP912 AlGaAs p-i-n diode. https://www.macom.com/products/product-detail/MA4AGBLP912. (visited: 2019-05-27).

[Mau05] Jürgen Maurer. *Strahlenoptisches Kanalmodell für die Fahrzeug-Fahrzeug-Funkkommunikation*. PhD thesis, 2005.

[ME16] B. Majumdar and K. P. Esselle. A single band beam scanning active phased array antenna. In *2016 International Conference on Electromagnetics in Advanced Applications (ICEAA)*, pages 832–835, Sept 2016.

[MMG17] R. Molina-Masegosa and J. Gozalvez. LTE-V for Sidelink 5G V2X Vehicular Communications: A New 5G Technology for Short-Range Vehicle-to-Everything Communications. *IEEE Vehicular Technology Magazine*, 12(4):30–39, Dec 2017.

[MRH+15] T. Mahler, L. Reichardt, C. Heine, M. Pauli, and T. Zwick. Channel Based Design of Systems with Multiple Antennas. *Progress In Electromagnetics Research B*, 64:63–81, 2015.

[NJDP12] V. Nguyen, M. Jeong, M. Dao, and S. Park. Four-port beam reconfigurable antenna array for pattern diversity system. *IET Microwaves, Antennas Propagation*, 6(10):1179–1186, July 2012.

[NTKHF15] N. Nguyen-Trong, T. Kaufmann, L. Hall, and C. Fumeaux. Analysis and Design of a Reconfigurable Antenna Based on Half-Mode Substrate-Integrated Cavity. *IEEE Transactions on Antennas and Propagation*, 63(8):3345–3353, Aug 2015.

[NTPHF17a] N. Nguyen-Trong, A. Piotrowski, L. Hall, and C. Fumeaux. A Frequency- and Polarization-Reconfigurable Circular Cavity Antenna. *IEEE Antennas and Wireless Propagation Letters*, 16:999–1002, 2017.

[NTPHF17b] N. Nguyen-Trong, A. Piotrowski, L. T. Hall, and C. Fumeaux. Concept of a beam-steerable cavity-fed antenna with magnetic-dipole coupling elements. In *2017 11th European Conference on Antennas and Propagation (EUCAP)*, pages 927–929, March 2017.

[NXP] NXP. BAP64-02Silicon PIN diode. https://www.nxp.com/docs/en/data-sheet/BAP64-02.pdf. (visited: 2019-05-27).

[OOA12] C. Oikonomopoulos-Zachos, T. Ould, and M. Arnold. Outdoor Channel Characterization of MIMO-LTE Antenna Configurations through Measurements. In *2012 IEEE 75th Vehicular Technology Conference (VTC Spring)*, pages 1–4, May 2012.

[ORPH+17] A. Ourir, K. Rachedi, D. . Phan-Huy, C. Leray, and J. de Rosny. Compact reconfigurable antenna with radiation pattern diversity for spatial modulation. In *2017 11th European Conference on Antennas and Propagation (EUCAP)*, pages 3038–3043, March 2017.

[OSM] OSM. OpenStreetMap Globe. http://osm-3d.org/. (visited: 2018-11-08).

[PK10] J. R. Panda and R. S. Kshetrimayum. A compact CPW-fed monopole antenna with an U-shaped slot for 5 GHz/6 GHz band-notched ultrawideband applications. In *2010 Indian Antenna Week: A Workshop on Advanced Antenna Technology*, pages 1–4, May 2010.

[Poz12] D. M. Pozar. *Microwave Engineering*. John Wiley & Sons, Hoboken, USA, 2012.

[PS09] John G. Proakis and Masoud Salehi. *Digital communications*. McGraw-Hilll higher education. McGraw-Hill, Boston [u.a.], 5. ed., internat. ed., [nachdr.] edition, 2009. ISBN in der Vorlage: 978-007-126378-8 und 007-126378-0.

[PSR+11] B. D. Pell, E. Sulic, W. S. T. Rowe, K. Ghorbani, and S. John. *Advancements in Automotive Antennas, New Trends and Developments in Automotive System Engineering*. InTech, Rijeka, Croatia, 2011.

[QGWL12] P. Qin, Y. J. Guo, A. R. Weily, and C. Liang. A Pattern Reconfigurable U-Slot Antenna and Its Applications in MIMO Systems. *IEEE Transactions on Antennas and Propagation*, 60(2):516–528, Feb 2012.

[Rei13] Lars Reichardt. *Methodik für den Entwurf von kapazitätsoptimierten Mehrantennensystemen am Fahrzeug*. PhD thesis, Karlsruhe, 2013.

[Rem] Remcom. Wireless InSite. https://www.remcom.com/ wireless-insite-em-propagation-software/. (visited: 2018-11-08).

[RMSZ13] L. Reichardt, T. Mahler, Y. L. Sit, and T. Zwick. Using a synthesis methodology for the design of automotive antenna systems. In *2013 7th European Conference on Antennas and Propagation (EuCAP)*, pages 1600–1604, April 2013.

[Sau99] Simon R. Saunders. *Antennas and propagation for wireless communication systems*. Wiley, Chichester, 1999.

[SMS06] D. Schurig, J. J. Mock, and D. R. Smith. Electric-field-coupled resonators for negative permittivity metamaterials. *Applied Physics Letters*, 88(4), 2006.

[SSA12] M. S. Sharawi, F. Sultan, and D. N. Aloi. An 8-Element Printed V-Shaped Circular Antenna Array for Power-Based Vehicular Localization. *IEEE Antennas and Wireless Propagation Letters*, 11:1133–1136, 2012.

[SUE12] Adnan Sondas, Mustafa H. B. Ucar, and Yunus E. Erdemli. Switchable loop-loaded printed dipole antenna with a balun/feed structure. *Microwave and Optical Technology Letters*, 54(1):76–79, 2012.

[SVM15] E. Safin, R. Valkonen, and D. Manteuffel. Reconfigurable LTE MIMO automotive antenna system based on the characteristic mode analysis. In *2015 9th European Conference on Antennas and Propagation (EuCAP)*, pages 1–3, April 2015.

[TC17] J. Tak and J. Choi. A flush-mounted monopolar patch antenna for UAV applications. *Microwave and Optical Technology Letters*, 59(5):1202–1207, 2017.

[VBA03] Rodney Vaughan and Jorgen Bach Andersen. *Channels, Propagation and Antennas for Mobile Communications*. The Institution of Engineering and Technology, Stevenage, 1. ed. edition, 2003.

[XYN09] T. Xia, S. Yang, and Z. Nie. Design of a Tapered Balun for Broadband Arrays With Closely Spaced Elements. *IEEE Antennas and Wireless Propagation Letters*, 8:1291–1294, 2009.

[YZC+16] D. Yang, H. Zeng, R. Chen, J. Qu, Y. Wen, and S. Liu. Four port compact multimode patch antenna system for vehicular application. In *2016 IEEE International Workshop on Electromagnetics: Applications and Student Innovation Competition (iWEM)*, pages 1–3, May 2016.

[ZAH14] Y. Zhou, R. S. Adve, and S. V. Hum. Design and Evaluation of Pattern Reconfigurable Antennas for MIMO Applications. *IEEE Transactions on Antennas and Propagation*, 62(3):1084–1092, March 2014.

[ZGM+18] G. B. Zhang, R. C. Gough, M. R. Moorefield, K. J. Cho, A. T. Ohta, and W. A. Shiroma. A Liquid-Metal Polarization-Pattern-Reconfigurable Dipole Antenna. *IEEE Antennas and Wireless Propagation Letters*, 17(1):50–53, Jan 2018.

[ZMR+13] A. Zohur, H. Mopidevi, D. Rodrigo, M. Unlu, L. Jofre, and B. A. Cetiner. RF MEMS Reconfigurable Two-Band Antenna. *IEEE Antennas and Wireless Propagation Letters*, 12:72–75, 2013.

[ZNTF18] S. N. M. Zainarry, N. Nguyen-Trong, and C. Fumeaux. A Frequency- and Pattern-Reconfigurable Two-Element Array Antenna. *IEEE Antennas and Wireless Propagation Letters*, 17(4):617–620, April 2018.

[ZWZ+15] Y. Zhang, K. Wei, Z. Zhang, Y. Li, and Z. Feng. A Compact Dual-Mode Metamaterial-Based Loop Antenna for Pattern Diversity. *IEEE Antennas and Wireless Propagation Letters*, 14:394–397, 2015.

[ZXX09] F. Zhu, J. Xu, and Q. Xu. Reduction of mutual coupling between closely-packed antenna elements using defected ground structure. *Electronics Letters*, 45(12):601–602, June 2009.

# Own Publications

## Journals

[1] M. Jalilvand, X. Li, J. Kowalewski, and T. Zwick. Broadband miniaturised bow-tie antenna for 3D microwave tomography. *Electronics Letters*, 50(4):244–246, February 2014.

[2] J. Kowalewski, J. Atuegwu, J. Mayer, T. Mahler, and T. Zwick. A Low-Profile Pattern Reconfigurable Antenna System for Automotive MIMO Applications. *Progress In Electromagnetics Research*, 161:41–55, 2018.

[3] J. Kowalewski, J. Eisenbeis, M. Tingulstad, Z. Kollar, and T. Zwick. Design Method for Capacity Enhancement of Pattern-Reconfigurable MIMO Vehicular Antennas. *IEEE Antennas and Wireless Propagation Letters*, pages 1–1, 2019.

[4] J. Kowalewski, T. Mahler, L. Reichardt, and T. Zwick. Shape Memory Alloy (SMA)-Based Pattern-Reconfigurable Antenna. *IEEE Antennas and Wireless Propagation Letters*, 12:1598–1601, 2013.

[5] T. Mahler, J. Kowalewski, B. Nub, C. Richt, J. Mayer, and T. Zwick. Channel Measurement Based Antenna Synthesis for Mobile Automotive MIMO Communication Systems. *Progress In Electromagnetics Research B*, 72:1–16, 2017.

## Conference Papers

[6] G. Artner, J. Kowalewski, J. Atuegwu, C. F. Mecklenbräuker, and T. Zwick. Electronically Steerable Parasitic Array Radiator Flush-Mounted for Automotive LTE. In *2019 13th European Conference on Antennas and Propagation (EuCAP)*, pages 1–5, March 2019.

[7] G. Artner, J. Kowalewski, C. F. Mecklenbräuker, and T. Zwick. Automotive pattern reconfigurable antennas concealed in a chassis cavity. In *2017 20th*

*International Symposium on Wireless Personal Multimedia Communications (WPMC)*, pages 265–270, Dec 2017.

[8] G. Artner, J. Kowalewski, C. F. Mecklenbräuker, and T. Zwick. Pattern reconfigurable antenna with four directions hidden in the vehicle roof. In *2017 International Workshop on Antenna Technology: Small Antennas, Innovative Structures, and Applications (iWAT)*, pages 82–85, March 2017.

[9] M. Harter, J. Kowalewski, L. Sit, M. Jalilvand, A. Ziroff, and T. Zwick. 3D Radar Surveillance of Humans for Safety and Security Applications. In *GeMiC 2014; German Microwave Conference*, pages 1–4, March 2014.

[10] M. Jalilvand, C. Vasanelli, J. Kowalewski, and T. Zwick. Implementation of Antenna Array Systems for Medical Imaging. In *GeMiC 2014; German Microwave Conference*, pages 1–4, March 2014.

[11] M. Jalilvand, C. Vasanelli, C. Wu, J. Kowalewski, and T. Zwick. On the evaluation of a proposed bowtie antenna for microwave tomography. In *The 8th European Conference on Antennas and Propagation (EuCAP 2014)*, pages 2790–2794, April 2014.

[12] J. Kowalewski, U. Dey, T. Mahler, and T. Zwick. Vivaldi antenna with improved directivity for medical applications. In *2015 German Microwave Conference*, pages 123–126, March 2015.

[13] J. Kowalewski, J. Eisenbeis, and T. Zwick. A Method for Determination of Reconfigurable Multiantenna Systems' Performance. In *2019 13th European Conference on Antennas and Propagation (EuCAP)*, pages 1–4, March 2019.

[14] J. Kowalewski, L. Keller-Bauer, T. Mahler, J. Mayer, and T. Zwick. Realization of a compact antenna with reconfigurable pattern for multiple antenna systems. In *2017 11th European Conference on Antennas and Propagation (EUCAP)*, pages 917–921, March 2017.

[15] J. Kowalewski, T. Mahler, C. Heine, and T. Zwick. Compact pattern reconfigurable LTE antenna. In *2014 International Workshop on Antenna Technology: Small Antennas, Novel EM Structures and Materials, and Applications (iWAT)*, pages 72–75, March 2014.

[16] J. Kowalewski, T. Mahler, J. Mayer, and T. Zwick. A miniaturized pattern reconfigurable antenna for automotive applications. In *2016 10th European Conference on Antennas and Propagation (EuCAP)*, pages 1–4, April 2016.

[17] J. Kowalewski, T. Mahler, L. Reichardt, and T. Zwick. Investigation of the influence of panoramic roof on mobile telephony antennas. In *The 8th*

*European Conference on Antennas and Propagation (EuCAP 2014)*, pages 1062–1066, April 2014.

[18] J. Kowalewski, T. Mahler, T. Schipper, and T. Zwick. Realization of a pattern reconfigurable antenna employing PIN diodes. In *2015 IEEE International Symposium on Antennas and Propagation USNC/URSI National Radio Science Meeting*, pages 2267–2268, July 2015.

[19] J. Kowalewski, J. Mayer, T. Mahler, and T. Zwick. A compact pattern reconfigurable antenna utilizing multiple monopoles. In *2016 International Workshop on Antenna Technology (iWAT)*, pages 1–4, Feb 2016.

[20] J. Kowalewski, J. Mayer, T. Mahler, and T. Zwick. A pattern reconfigurable microstrip antenna for future automotive applications. In *12th European Conference on Antennas and Propagation (EuCAP 2018)*, pages 1–5, April 2018.

[21] J. Kowalewski, J. Mayer, T. Mahler, and T. Zwick. Evaluation of a wideband pattern-reconfigurable multiple antenna system for vehicular applications. In *12th European Conference on Antennas and Propagation (EuCAP 2018)*, pages 1–5, April 2018.

[22] J. Kowalewski, I. Mehinovic, S. Abadpour, J. Mayer, and T. Zwick. A pattern reconfigurable antenna system for automotive MIMO applications. In *2018 11th German Microwave Conference (GeMiC)*, pages 29–32, March 2018.

[23] J. Kowalewski, S. Peukert, T. Mahler, J. Mayer, and T. Zwick. A trident like antenna with reconfigurable patterns for automotive applications. In *2017 International Workshop on Antenna Technology: Small Antennas, Innovative Structures, and Applications (iWAT)*, pages 96–99, March 2017.

[24] T. Mahler, T. Deletoille, J. Frey, J. Kowalewski, and T. Zwick. Applying antenna synthesis methods on a path based MIMO channel model for verification. In *2017 47th European Microwave Conference (EuMC)*, pages 1349–1352, Oct 2017.

[25] T. Mahler, T. Deletoille, J. Frey, J. Kowalewski, and T. Zwick. Applying antenna synthesis methods on a path based MIMO channel model for verification. In *2017 European Radar Conference (EURAD)*, pages 501–504, Oct 2017.

[26] T. Mahler, J. Kowalewski, L. Reichardt, and T. Zwick. Realization of a synthesized compact automotive roof-top LTE antenna. In *2013 IEEE*

*Antennas and Propagation Society International Symposium (APSURSI)*, pages 2073–2074, July 2013.

[27] T. Mahler, J. Kowalewski, T. Schipper, and T. Zwick. Channel capacity determination of a pattern reconfigurable automotive roof-top LTE antenna. In *2014 Asia-Pacific Microwave Conference*, pages 974–976, Nov 2014.

[28] T. Mahler, J. Kowalewski, T. Schipper, and T. Zwick. A pattern reconfigurable automotive LTE antenna employing synthesized radiation patterns. In *2015 IEEE MTT-S International Conference on Microwaves for Intelligent Mobility (ICMIM)*, pages 1–4, April 2015.

[29] T. Mahler, C. Richt, L. Bell, M. Pauli, J. Kowalewski, and T. Zwick. Antenna synthesis for SIMO and MISO systems with optimality to arbitrary quantiles of the channel capacity. In *2018 IEEE Radio and Wireless Symposium (RWS)*, pages 35–37, Jan 2018.

[30] J. Mayer, J. Kowalewski, M. Vieweger, J. Eisenbeis, T. Mahler, and T. Zwick. Reflectarray with split ring resonators at 83.5 GHz. In *2017 47th European Microwave Conference (EuMC)*, pages 1277–1280, Oct 2017.

[31] J. Mayer, B. Nuss, J. Kowalewski, and T. Zwick. Angular Resolution Estimation for Conformal Radar Antenna Arrays. In *2018 IEEE MTT-S International Conference on Microwaves for Intelligent Mobility (ICMIM)*, pages 1–4, April 2018.

[32] L. Reichardt, J. Kowalewski, L. Zwirello, and T. Zwick. Compact, Teflon embedded, dual-polarized ultra wideband (UWB) antenna. In *Proceedings of the 2012 IEEE International Symposium on Antennas and Propagation*, pages 1–2, July 2012.

[33] T. Schipper, J. Schlichenmaier, D. Ahbe, T. Mahler, J. Kowalewski, and T. Zwick. A simulator for multi-user automotive radar scenarios. In *2015 IEEE MTT-S International Conference on Microwaves for Intelligent Mobility (ICMIM)*, pages 1–4, April 2015.

**Forschungsberichte aus dem Institut für Höchstfrequenztechnik und Elektronik (IHE) der Universität Karlsruhe (TH)**
(ISSN 0942-2935)

---

Herausgeber: Prof. Dr.-Ing. Dr. h.c. Dr.-Ing. E.h. mult. Werner Wiesbeck

Die Bände 1 (1992) bis 55 (2008) der Schriftenreihe können über das Institut Hochfrequenztechnik und Elektronik bestellt werden (https://www.ihe.kit.edu).

Fortführung als:

**Karlsruher Forschungsberichte aus dem Institut für Hochfrequenztechnik und Elektronik**
(ISSN 1868-4696)

---

Herausgeber: Prof. Dr.-Ing. Thomas Zwick

Band 55 Sandra Knörzer
**Funkkanalmodellierung für OFDM-Kommunikationssysteme bei Hochgeschwindigkeitszügen (2009)**
ISBN 978-3-86644-361-7

Band 56 Thomas Fügen
**Richtungsaufgelöste Kanalmodellierung und Systemstudien für Mehrantennensysteme in urbanen Gebieten (2009)**
ISBN 978-3-86644-420-1

Band 57 Elena Pancera
**Strategies for Time Domain Characterization of UWB Components and Systems (2009)**
ISBN 978-3-86644-417-1

Band 58 Jens Timmermann
**Systemanalyse und Optimierung der Ultrabreitband-Übertragung (2010)**
ISBN 978-3-86644-460-7

---

Band 59 Juan Pontes
**Analysis and Design of Multiple Element Antennas for Urban Communication (2010)**
ISBN 978-3-86644-513-0

Band 60 Andreas Lambrecht
**True-Time-Delay Beamforming für ultrabreitbandige Systeme hoher Leistung (2010)**
ISBN 978-3-86644-522-2

Band 61 Grzegorz Adamiuk
**Methoden zur Realisierung von dual-orthogonal, linear polarisierten Antennen für die UWB-Technik (2010)**
ISBN 978-3-86644-573-4

Band 62 Jutta Kühn
**AlGaN/GaN-HEMT Power Amplifiers with Optimized Power-Added Efficiency for X-Band Applications (2011)**
ISBN 978-3-86644-615-1

Band 63 Małgorzata Janson
**Hybride Funkkanalmodellierung für ultrabreitbandige MIMO-Systeme (2011)**
ISBN 978-3-86644-639-7

Band 64 Mario Pauli
**Dekontaminierung verseuchter Böden durch Mikrowellenheizung (2011)**
ISBN 978-3-86644-696-0

Band 65 Thorsten Kayser
**Feldtheoretische Modellierung der Materialprozessierung mit Mikrowellen im Durchlaufbetrieb (2011)**
ISBN 978-3-86644-719-6

Band 66 Christian Andreas Sturm
**Gemeinsame Realisierung von Radar-Sensorik und Funkkommunikation mit OFDM-Signalen (2012)**
ISBN 978-3-86644-879-7

---

Band 67 Huaming Wu
**Motion Compensation for Near-Range Synthetic Aperture Radar Applications (2012)**
ISBN 978-3-86644-906-0

Band 68 Friederike Brendel
**Millimeter-Wave Radio-over-Fiber Links based on Mode-Locked Laser Diodes (2013)**
ISBN 978-3-86644-986-2

Band 69 Lars Reichardt
**Methodik für den Entwurf von kapazitätsoptimierten Mehrantennensystemen am Fahrzeug (2013)**
ISBN 978-3-7315-0047-6

Band 70 Stefan Beer
**Methoden und Techniken zur Integration von 122 GHz Antennen in miniaturisierte Radarsensoren (2013)**
ISBN 978-3-7315-0051-3

Band 71 Łukasz Zwirełło
**Realization Limits of Impulse-Radio UWB Indoor Localization Systems (2013)**
ISBN 978-3-7315-0114-5

Band 72 Xuyang Li
**Body Matched Antennas for Microwave Medical Applications (2014)**
ISBN 978-3-7315-0147-3

Band 73 Sebastian Diebold
**Transistor- und Leitungsmodellierung zum Entwurf von monolithisch integrierten Leistungsverstärkern für den hohen Millimeterwellen-Frequenzbereich (2014)**
ISBN 978-3-7315-0161-9

Band 74 Christian Rusch
**Integrierte, planare Leckwellenantennen für 3D-Millimeterwellen-Radarsysteme basierend auf dem holografischen Prinzip (2014)**
ISBN 978-3-7315-0234-0

---

Band 75 Marlene Harter
**Dreidimensional bildgebendes Radarsystem mit digitaler Strahlformung für industrielle Anwendungen (2014)**
ISBN 978-3-7315-0249-4

Band 76 Michael A. Baldauf
**Abhängigkeit der Exposition von der Zellgröße beim Mobilfunk unter Gewährleistung der Versorgung (2015)**
ISBN 978-3-7315-0308-8

Band 77 Alicja Ossowska
**Highly Resolved Synthetic Aperture Radar with Beam Steering (2015)**
ISBN 978-3-7315-0315-6

Band 78 Małgorzata Dominika Brzeska
**RF Modelling and Characterization of Tyre Pressure Sensors and Vehicle Access Systems (2015)**
ISBN 978-3-7315-0348-4

Band 79 Ulrich Lewark
**Aktive Frequenzvervielfacher zur Signalerzeugung im Millimeter- und Submillimeterwellen Frequenzbereich (2015)**
ISBN 978-3-7315-0354-5

Band 80 Kai-Philipp Walter Pahl
**Distributed Transformers for Broadband Monolithic Millimeter-Wave Integrated Power Amplifiers (2015)**
ISBN 978-3-7315-0409-2

Band 81 Serdal Ayhan
**Hochgenaue radarbasierte Abstandsmessung mit geführter Wellenausbreitung (2016)**
ISBN 978-3-7315-0433-7

Band 82 Yoke Leen Sit
**MIMO OFDM Radar-Communication System with Mutual Interference Cancellation (2017)**
ISBN 978-3-7315-0599-0

Band 83 Steffen Scherr
**FMCW-Radarsignalverarbeitung zur Entfernungsmessung mit hoher Genauigkeit (2017)**
ISBN 978-3-7315-0607-2

Band 84 Tom Schipper
**Modellbasierte Analyse des Interferenzverhaltens von Kfz-Radaren (2017)**
ISBN 978-3-7315-0639-3

Band 85 Malyhe Jalilvand
**Application-Specific Broadband Antennas for Microwave Medical Imaging (2017)**
ISBN 978-3-7315-0664-5

Band 86 Benjamin Göttel
**Millimeterwellen On-Chip Antennensysteme für die Integration in SoC Applikationen (2017)**
ISBN 978-3-7315-0667-6

Band 87 Christian Arnold
**Im Orbit einstellbare Ausgangsfilter und -multiplexer (2017)**
ISBN 978-3-7315-0722-2

Band 88 Tobias Mahler
**Synthese kapazitätsoptimierter Antennensysteme mit messtechnischer Verifikation (2018)**
ISBN 978-3-7315-0737-6

Band 89 Daniel Müller
**RF Probe-Induced On-Wafer Measurement Errors in the Millimeter-Wave Frequency Range (2018)**
ISBN 978-3-7315-0822-9

Band 90 Tristan Visentin
**Polarimetric Radar for Automotive Applications (2019)**
ISBN 978-3-7315-0888-5

Band 91 Christian von Vangerow
**Entwurf und Modellierung von Breitbandverstärkern mit variablem Gewinn in SiGe BiCMOS Technologien (2019)**
ISBN 978-3-7315-0910-3

Band 92 Mekdes Girma
**Concepts for Short Range Millimeter-wave Miniaturized Radar Systems with Built-in Self-Test (2019)**
ISBN 978-3-7315-0938-7

Band 93 Akanksha Bhutani
**Low Temperature Co-fired Ceramics for System-in-Package Applications at 122 GHz (2019)**
ISBN 978-3-7315-0945-5

Band 94 Jochen Schäfer
**Oberflächenwellenerzeuger für Millimeterwellen-Leckwellenantennen (2019)**
ISBN 978-3-7315-0962-2

Band 95 Jerzy Kowalewski
**Capacity Enhancement by Pattern-Reconfigurable Multiple Antenna Systems in Vehicular Applications (2020)**
ISBN 978-3-7315-0997-4